SAVING CREATION

SAVING CREATION

Nature and Faith in the Life of

HOLMES ROLSTON III

Christopher J. Preston

Trinity University Press
San Antonio

Published by Trinity University Press
San Antonio, Texas 78212
www.trinity.edu/tupress

Jacket design by David Drummond
Book design by BookMatters, Berkeley

♾ The paper used in this publication meets the minimum
requirements of the American National Standard for
Information Sciences—Permanence of Paper for Printed
Library Materials, ANSI Z39.48-1992.

Library of Congress Cataloging-in-Publication Data

Preston, Christopher J. (Christopher James), 1968–
Saving creation : nature and faith in the life of Holmes
Rolston III / Christopher J. Preston.
p. cm.
Includes bibliographical references.
ISBN 978-1-59534-050-4 (hardcover : alk. paper)
1. Rolston, Holmes, 1932– 2. Rolston, Holmes,
1932– —Religion. 3.Rolston, Holmes, 1932——Philosophy.
4. Environmentalists—United States—Biography.
5. Naturalists—United States—Biography. 6. Philosophers—
United States—Biography. 7. Environmental ethics.
8. Religion and science. 9. Nature—Religious aspects. I. Title.
GE56.R64P74 2009
333.72092—dc22
[B] 2008038996

13 12 11 10 09 C 5 4 3 2 1

Contents

Acknowledgments

I wish to thank the following friends, family, and acquaintances for giving me assistance and invaluable guidance about how to best tell this story: Larry Godwin, Peter Stark, Deborah Slicer, Sandy Woodson, Jeff Stolle, Bruce Coull, Louise Westling, William Bailey, Lisa Sideris, Robin and Wendy Preston, Lisa Miller, Bob Crumby, Malcolm Brownlee, Bill Forbes, Shonny Vander Vliet, Perry Biddle, Ron Williams, Fred Johnson, Phillip Cafaro, Michael Nelson, Bernie Rollin, Ned Hettinger, and Andrew Light. Together they added important details and managed to temper considerably my clumsiness with words.

The John Templeton Foundation supported ten months of the initial research and writing of this book. I appreciate its directors' confidence in me and the complete freedom they gave me to tell the story.

Thanks also to Barbara Ras, Sarah Nawrocki, and Claudia Guerra at Trinity University Press for their help, enthusiasm, and support.

Great appreciation goes to Holmes and Jane Rolston for their assistance at every step of the way in the writing of this book. Having your story told by somebody else must be an unnerving process. Holmes Rolston provided all the information I requested, opened his archives, scrapbooks, and photo albums to my searching, gave hours of formal and informal interviews, and followed up every e-mail I sent him. Jane also showed remarkable kindness and tolerance of my intrusions into her life and responded to numerous of my requests for assistance. At the same time as giving me abundant material to work with, both Holmes and Jane stood back admirably from determining the content of the book. I only hope I have done justice in what follows to a remarkable life and a remarkable partnership.

Special thanks go to Lisa Miller, who has been beside me from the very beginning of this project. Nothing more can be asked of a partner than honest advice, thoughtful support, and abundant love throughout a challenging process. Lisa, as ever, provided all of these, topped with uplifting smiles, words of encouragement, and exhausting Montana hikes and bike rides when I needed them most. My thanks and my love.

Finally, the nine weeks my mother spent in the hospital during the final period of the writing of this book taught me a great deal about suffering, about strength, and about character. Members of my family have faced substantial challenges over the last few years. As with the pasqueflower Rolston seeks in the Colorado high country, we search for beauty emerging from the struggle and sometimes we find it.

Introduction

Holmes Rolston III sat at his desk gazing at the fields and woods of the Valley of Virginia and reflecting on the conversation just ended with two of the elders from High Point Presbyterian Church. It was the fall of 1965. Like his father and grandfather before him, the thirty-two-year-old pastor had worked hard tending to the spiritual needs of his congregation. He knew he was pretty good at what he did. He had come to the pastor's life from a family steeped in church tradition. In addition to the family background, he had gained a top-of-the-line seminary education in Virginia and in Scotland, the birthplace of Presbyterianism. The young cleric was comfortable and mostly happy in his work. He and his devoted wife, Jane, felt they belonged in a parish in their native Virginia. The Rolstons seemed a near-perfect fit at High Point, yet it was clear that just beneath the surface something had been going badly wrong.

The news that Rolston's church's elders were to petition the regional presbytery to remove him from his duties did not come as a complete surprise to the hard-working young minister. He had sensed the growing discontent of his congregation and heard the whispers circulating after church on Sunday. He knew many of the parishioners were uneasy about him and about the way he delivered his message. The atmosphere in his church had been deteriorating for months and it was clear to everybody that something would have to give.

Though the knowledge he was about to be fired from a job that both his father and grandfather had performed with distinction before him certainly stung a little, something inside of Rolston agreed with the elders' decision that it was time for him to go. He had sensed there was something else, something unique, he needed to achieve in his life. As he looked out of his window at the gently forested hills and started to adjust his mind to the fact he would be leaving his rural Virginia church, he felt a faint flicker of excitement at the prospect of what might lie ahead.

In May of 2003, the Duke of Edinburgh awarded Holmes Rolston III the Templeton Prize for discoveries in science and religion at Buckingham Palace in London. The prize is the world's richest single award in recognition of an individual's lifework. Former winners of the Templeton Prize have included Mother Teresa of Calcutta and Soviet dissident Aleksandr Solzhenitsyn. Six years previously Rolston had received a similarly high level of recognition when he was selected to deliver the 1997 Gifford Lectures at the University of Edinburgh. This annual series, established in memory of Scottish judge Lord Adam Gifford in 1888, is a forum for the world's most influential thinkers on theology and nature. Former Gifford lecturers have included American philosopher William James and Swiss missionary Albert Schweitzer. Between being fired from High Point in 1965 and the new millennium, Rolston had clearly achieved something of startling intellectual significance. This book is an attempt to chronicle that achievement.

From the birth of modern science in the seventeenth century to the mapping of the human genome in the twenty-first, the puzzle of how to set the natural sciences comfortably alongside Christian theology has never been entirely resolved. A world explained by physical and chemical laws sits awkwardly alongside a world explained by the plan of a divine creator. Evolutionary science in general, and the evolution of humankind in particular, often seems at odds with the Christian idea of humans created in the image of God. Charles Darwin spent a lifetime wrestling with the conflict and never reached a settled state of mind on the topic. The ever-increasing breadth and depth of knowledge about the natural world since Darwin first published his ideas one hundred and fifty years ago has made the puzzle only harder to solve.

From his earliest education in Virginia and North Carolina, Holmes Rolston III had always been a pretty good scientist. In time off from his clerical duties, he had also taught himself to be an accomplished naturalist. It was this competence in the sciences that created the problem in his ministry. Many of the congregation at High Point were intimidated and annoyed by the science their "modern reverend" studied in his spare time. His evolutionary account of origins clashed with the creation stories told in their Bible. It was not clear to them what sort of spiritual guide this modern and educated young minister could be.

The young pastor knew instinctively that his congregation's rejection of modern science spelled trouble. This trouble was not limited to his congregation, it extended to the church as a whole if it failed to reconcile Christian orthodoxy with modern scientific theory. He knew that evolutionary theory and Christianity needed to find a more comfortable way to coexist. By 1965, Rolston was starting to sense that articulating this coexistence would be one of his life's most important tasks.

On top of Christianity's struggle to accept evolution, there was a second consideration pushing Rolston away from his ministry at High Point Presbyterian. His love of the rural Virginia countryside in which he and his ancestors had spent their lives was slowly switching over to a feeling of despair at its destruction. An ever-expanding cycle of logging, mining, road building, and housing construction was tearing the fabric of the landscape apart before his eyes. The economic expansion of the postwar period was finally causing some Americans to wake up to its effects on the natural world.

Rolston knew that this large-scale destruction stemmed in some measure from a failure to recognize nature's moral and religious significance. He also knew that Christianity's strained relationship with the biological sciences was a contributing factor. Together with a few other progressive minds scattered across the country in the 1960s, Rolston saw the need for a dramatic philosophical and theological shift in the way people thought about the natural world. Articulating this shift was something Rolston was starting to see as his second life task.

Rolston's worries about his congregation at High Point turned out to be largely prophetic. Questions of how to think about the relationships between nature, morality, and divinity still cut a wide swathe through contemporary life. Close to fifty years after Rolston started worrying about these issues, the tension between evolutionary science and theology remains acute. A century and a half after the publication of Darwin's *Origin of Species* more than 40 percent of Americans—ignoring all scientific evidence to the contrary—still believe humans were created by God fewer than ten thousand years ago. Christianity and modern biology continue to tussle in areas that range from the ethics of stem cell research, to the genetic modification of organisms, to the teaching of intelligent design. Biology and theology are still nowhere near to being at peace while ecological systems continue to come apart under the pressures placed on them by humans. In many of these debates, society appears to be stuck at a crossroads between approaches claiming to be secular and scientific and approaches claiming to be traditionally Christian. Politicians use this ideological fault line to pit constituents against each other. More often than not the scientific and religious frameworks compete for priority *within* individuals as well as *between* them. The tension is an ancient one but today's technology is making the stress points more acute. Sometimes the chasm between biology and theology seems impossible to bridge.

Rolston has spent his entire intellectual life within this contested arena. While the body of work that eventually earned Rolston international recognition represents progress and discovery at the cutting edge of academic philosophy and theology, the motivation for this book does not come from academics. The tensions Rolston speaks to in his work are central tensions in human life, found at the place where God, nature, and humanity meet. Some of the most trenchant and complex puzzles for philosophers and theologians lie at this intersection, along with some of the most basic and fundamental questions of everyday life. What is our place on earth? Why do we suffer? How should we understand right and wrong on a living planet? Is care for the earth a matter of human preference, divine will, or both? What exactly is involved in saving creation? Where on earth—literally—do we go for answers to these questions? It is this mixture of the frighteningly complex with the absolutely fundamental that makes the work of this former Virginia pastor so vital today.

While the territory is complex, Rolston's journey through it can be told relatively simply. It is possible to approach the ideas for which Rolston has won such acclaim by following a careful biographical track through his life. "Human life," Rolston once wrote, "will not be a disembodied reason but a person organic in history. Character always takes a narrative form." This emphasis on personal history is especially appropriate in an account of Rolston's own journey. Often more interesting and powerful than any of his intellectual influences are those of the family members he looked up to and the distinctive landscapes in which he dwelled. His childhood in Virginia, his move to the mountains of Colorado, and Will Long, his Alabama grandfather, have shaped Rolston's thinking as much as the theology of Karl Barth or the philosophy of Aldo Leopold.

In this book, I attempt to chart the forces operating on Rolston's mind at different times in his life, beginning with childhood and continuing through his years as a pastor, a professional philosopher, and an environmentalist. I give the details of the cultural and natural environments in which he was raised and the lingering influence of the people and the landscapes that moved him. This strategy pries apart the sequence of ideas that by story's end come together into a complex but insightful intellectual whole. Telling the tale this way puts a human face on the adventure, connecting it to experiences imaginable by those who might be reluctant to enter the intellectual territory without a guide.

In the late 1960s, soon after receiving the letter from the elders at his rural Virginia church, Rolston embarked on an intellectual journey whose conclusion lies far beyond what one individual can hope to complete. Close to half a century later, few people can claim to have made more genuine progress in reconciling the central areas of philosophical and theological dissonance than Holmes Rolston III. His work presents invaluable insights about how science, religion, and secular environmentalism can find a harmonious relationship, something desperately needed today. Contemporary environmental thinking cannot be fully understood without knowing Rolston's work. Contemporary Christianity will not make sense without it. His story provides clear signposts through some of the murkiest landscapes that the human intellect must travel.

PART I

Southern Grounding

1

Shenandoah Valley Childhood

The logic of life is both biography and geography. The etymology of "biography" is to graph a life; the etymology of "geography" is to graph that life on Earth. . . . Biology requires geography. Life is always taking a journey through time and place. [*1998*]

FEW DOUBT THAT THE WAYS we think about the world are shaped by the barrage of instructions we endure from all manner of teachers and elders on the journey into adulthood. Fewer still take the trouble to note that the landscape itself is the grounding substrate upon which all of these cultural forces rest. Living beings enjoy what Rolston likes to call a "storied residence" in some environment. The philosopher's own storied residence begins firmly in the geography of Virginia's Shenandoah Valley.

The Shenandoah Valley is a gentle, diagonal crease in the landscape lying a hundred and fifty miles inland from the mid-Atlantic seaboard. The Blue Ridge Mountains bound the valley to the east and the Alleghenies mark its limit to the west. Over the crest of the Blue Ridge, the central and eastern portions of Virginia spread out across the Piedmont toward the coastal plain and the Atlantic Ocean. These flat and fertile landscapes supported the large tobacco plantations once responsible for Virginia's colonial wealth. The valleys of the Shenandoah and the highlands of the Cumberland Plateau to the west remained much poorer because of the harder topographies, the colder climates, and the barriers to transport created by ridges and valleys formed during the Alleghenian orogeny some 350 million years ago.

The roots of the word "Shenandoah" have long been lost to the past. The origin account the locals prefer talks of an indigenous word that means "Clear-Eyed-Daughter-of-the-Stars." The sparkle of the collected Appalachian rains coursing over beds of limestone in the valley bottoms suggested to the Shenandoah's first indigenous residents a landscape with its origins in the spirit world. The Scottish-Irish Presbyterians colonizing the valley in the eighteenth century were similarly convinced their chosen home had been blessed by the divine. Spring temperatures beginning in early March, summer afternoon skies rent by nourishing rainstorms, and a rich soil capable of germinating just about any seed you cared to throw at it created a landscape that appeared heaven-sent when contrasted with the bleak Scottish highlands from which they had come.

Holmes Rolston III was born on an unusually chilly mid-November night in 1932. An early winter storm had dusted the tops of the barns with a light snow. His father, the pastor at Bethesda Presbyterian Church in Rockbridge Baths, had been getting ready for evening service when word came his wife was in labor in nearby Staunton. He quickly canceled the service and jumped into his Ford Model T. The pastor sped the thirty-four miles into town, sliding to a stop in front of the hospital shortly before 8 p.m. By the time he had raced up the steps to his wife's bedside, the contractions seizing Mary Long Rolston's abdomen signaled the imminent arrival of the couple's first child.

Rolston drew his first breath in a building now part of Mary Baldwin College in Staunton, Virginia. Theron Rolston, the pastor's brother, was the attending physician. The newborn and his mother spent the next six days recuperating at a tidy red brick house on New Street not far from the hospital. The owner of the house, the baby's widowed grandmother, assisted by preparing hot baths, washing diapers and linens, and cooking for her two charges. A few days before the end of November, the pastor drove the newborn and his mother back to Rockbridge Baths and the baby was carried into the building that would be home for the first nine years of his life. Once inside, he was set down on a simple wooden cradle not far from a blazing wood stove tended by Miss Ida Whitsell, the practical nurse the Rolstons had hired to help out in the early months. For the next few days, Holmes Rolston III nursed, cried, and slept as thin patches of snow melted on the ground outside.

The house in which the baby lay was a white, two-story wooden manse built at the turn of the nineteenth century. It stood just a few hundred feet to the east of the brick church where Holmes Rolston II preached his Sunday services. The living room windows looked out upon a near-perfectly composed image of Virginia's rural landscape. The Maury River ran by the church less than a quarter mile to the west and a hundred feet below. A rolling tapestry of small fields punctuated by creeks and woodlands covered the valley bottoms in all directions. Fields that in summer bristled with corn and wheat alternated with cow and sheep pastures, creating a soft, pastoral scene. In the hedgerows and woodlands, signs of deer, wild turkey, and bobcat announced that, underneath its Anglo-Saxon surface, this was indeed a landscape of the New World.

The manse and surrounding garden bore many of the hallmarks of rural life in 1930s America. There was no electricity in the house and the heat came from wood. A cistern nearby supplied the family's water. The Rolstons' home was one of only two in the area with rudimentary plumbing, but the system never worked very well and the family used the outhouse most of the time. Fifty feet from the back door rose a tall woodpile, most of it in the open air but enough under cover for kindling in rainy weather. Axes, wedges, a chopping block, and a crosscut saw stood nearby. Chickens scratched in the surrounding dirt. On a hill to the east, his father had built a backup water supply, connected to the house through a gravity-fed pipe. Rolston took numerous soggy trips in the years ahead up "Bunkum Hill," as the locals called it, to poke the water pipe clear after a rainstorm. Life was simple but comfortable. His parents assured him they were blessed to live within the Shenandoah.

When old enough to dream of his own adventures, Rolston gazed from the yard toward Jump and Hogback Mountains on the skyline to the northwest. Each of these rounded Appalachian peaks rose to just over three thousand forested feet, bringing deep evening shadows to the valleys carved beneath their flanks. The two mountains were separated by the upper reaches of the Maury, part of a river system containing what are thought to be some of the oldest rivers in the world, their flow occasionally cutting at right angles to the direction of the mountain ridges. These "water gaps" where the rivers crossed the mountains kept the passes low and made it possible to wade into a sizable river to fish, while behind each shoulder stood the highest peaks in the county.

The gap between Jump and Hogback was known as Goshen Pass. Rolston recalls many happy hours fishing and hiking there with his father. At Goshen Pass he discovered a broad array of flora and fauna challenging his senses. Mountain laurel and rhododendron shrubs were shaded by hickory and chestnut oak. The forested canopy was a haven for catbirds, yellow-rumped warblers, and indigo buntings. Beetles, slugs, and salamanders thickly populated the leafy forest floor. An average annual rainfall of thirty-eight inches, much of it falling during crashing afternoon thunderstorms, kept the ground damp and ensured that the woodlands harbored a rich community of ferns and mosses.

When looking back on those early years, Rolston recalls the Shenandoah Valley as a teeming biotic ark for a barefooted young boy to explore. He spent long childhood days investigating the Appalachian ecology that enveloped him, finding little reason to doubt his parents' claim that the Lord had indeed blessed the landscapes in which they dwelled. From the age of five Rolston wandered alone down to Hays Creek, half a mile from home, with his mother's permission. The young explorer would set out in search of the wildlife that left its signs along the banks. He stuck his nose close to the mud and examined the footprints left by otters and raccoons. He scrutinized dragonflies on azalea bushes and beetles plodding slowly between tree roots. He fished the creeks for hours at a time, hurrying proudly back to the manse to show his parents what he caught. In the family kitchen, his mother would often have to unhook the brook trout or smallmouth bass from the end of her son's fishing line. Though he was brave enough to explore the fields and woods alone, the thought of trying to grasp the flopping and slippery creatures he had caught remained for now beyond the bounds of Rolston's courage.

Rolston was mostly comfortable around water. He and his two younger sisters, Mary Jacqueline and Julia, learned to swim in the Maury just below their home. The river provided welcome respite from the humidity of Virginia's long summer afternoons. But the playfulness did come with an edge. Hays Creek had earlier come close to bringing an abrupt end to the young Virginian's life. One wet spring afternoon when Rolston was still an infant, his father had tried to drive the family car across a bridge washed with floodwaters. The motor drowned and the car spluttered to a halt midstream. With wife and child marooned, the pastor waded off for help. A team of horses fetched from a nearby farm pulled the Model T, an alarmed Mrs. Rolston, and a screaming infant from the deluge. Passersby in later years would see a freckled young boy standing transfixed on the banks of the Maury watching the hydraulic ram pumping water to nearby farmhouses or staring at the single flickering electric lightbulb, powered by a generator attached to the mill wheel a quarter mile upriver, the only such light in the community. The curious child already sensed there were mysterious powers lurking beneath nature's surface.

Other lasting lessons about nature's productive power were available closer to home. In the yard surrounding the house, Rolston's father tended a large vegetable patch. After finishing his pastoral duties on weekdays, his father would spend part of the afternoon at work among the greens, weeding and turning the compost. Rolston learned from his parents how to pick tomatoes, shell beans, and husk corn. In late summer and early fall, the family would gather together to can the homegrown produce for winter. Like most other children in the valley in the 1930s, Holmes Rolston III grew up with the idea that food came from the ground not from the grocery store.

His father told him that working the land created character. Farmers "did not write bad checks for seed corn," the older Rolston said. Once, years later, worried that the gardener living at the trailhead had seen him leave the car unlocked, Rolston asked his father if he should run back and lock the vehicle. His father told him to relax about the car. Had he seen how the man kept his tomatoes?

In addition to the vegetables, Rolston's father also grew fruit. The pastor and Uncle Theron owned a five-acre apple orchard handed down to them through the family. Rolston fondly remembers the family trips to the orchard. His father was immensely proud of his Stayman-Winesaps. Father and son would scrape droppings from the chicken run at home and spread the manure around the base of the trees. At harvest time, with the girls' help, they picked the boughs clean, placing the apples in baskets delivered to a cold storage warehouse in Staunton. Every few weeks through the fall someone visited the warehouse to restock the cellar at home. The orchard was a productive one, and there was rarely a shortage of apples at the manse, with plenty more to give away. The family always spent a couple of days in late fall pressing some of the apples to make cider. They placed the cider in a big barrel in the basement with the door left ajar so the children could help themselves from the fragrant subterranean storehouse.

Without the benefit of electricity, Rolston's young body lived the passage from humid summer into the snows of winter. In January, residents of the valley hiked down to the Maury to cut ice, which they placed in their cold cellars to keep the stores cool. Rolston recalls frequent winter anxiety about whether the temperatures would get cold enough for good ice to form. He remembers putting river ice chips in early summer tea when the temperatures started to climb again. The ice was not always the cleanest, but it was widely believed by the locals that freezing the water purified it. In the days before industrial agriculture hit the valley, evidence against this folk wisdom was scarce. Valley residents drank from the springs and creeks without a thought.

The Shenandoah Valley provided Rolston with his first instruction on the fullness and fertility of the natural world. He was gripped by the miniature worlds of insects and amphibians he encountered in virtually any spot he stopped to explore. The valley had the right kind of natural richness to stimulate a curious young mind. By some standards, it was not an easy place to live. The winters could be cold and the summer humidity was stifling. Poverty took its toll. More than a few parishioners in his father's church died prematurely from disease or from alcoholism. In a letter from this period to his mother-in-law on the gulf coast, Rolston's father had written: "Times are mighty tight in Virginia and I suppose are even tighter in the far South. But we still have enough for the necessities and a little bit over." Like most of the Presbyterians who dwelled in the valley, the Rolston family was convinced this was a chosen place. His father, raised in the Shenandoah Valley himself, was deeply attached to the landscapes of home and passed on this sense of pride to his children, telling stories of family adventure in every district they visited. The verses from Psalms his father preached in Sunday service seemed as if they could easily have been written with the Shenandoah Valley in mind:

> You visitest the earth and waterest it,
> You greatly enrich it . . .
> You waterest its furrow abundantly,
> settling its ridges,
> Softening it with showers,
> and blessing its growth . . .
> The pastures of the wilderness drip,
> the hills gird themselves with joy,
> the meadows clothe themselves with flocks,
> the valleys deck themselves with grain,
> they shout and sing together for joy. (Psalm 65: 9–13)

For the rest of his life, even when living half a continent to the west in the Rocky Mountains, Rolston would still claim that in April and May there was no place on earth he would rather be than in Virginia's Shenandoah Valley.

The Shenandoah was not the only landscape that had a formative influence on Rolston's childhood. His mother had grown up on a farm in Perry County, in central Alabama. She had met Rolston's father at the local Presbyterian church, when he had moved to Alabama to teach mathematics at the Marion Military Institute after graduating from Washington and Lee. Mary Long Rolston took her children by train back to the family farm each summer to spend six weeks with her parents and sisters. Her twin sister, Willie Lee, had three children of her own and another sister, Carrie, had three more. There could be up to nine young cousins together on the Alabama farm. Rolston's father usually stayed the first few weeks back at Bethesda Presbyterian, grateful for some quiet time to gather his thoughts and write his sermons in peace. He would typically join his wife and children on the farm for a couple of weeks before vacation's end.

The Alabama farm was located on the fertile Black Belt soil near a swampy area on Bogue Chitto Creek. The drier land to the north of the creek was bisected by a railroad along which a locomotive rattled twice each day. A bluff overlooking the creek housed a small but impressive southern hardwood forest thick with pignut hickory, swamp bay, and black tupelo. At the bottom of the bluff, Will Long had built a pond where he taught his grandchildren to fish.

The children spent their summers immersed in the steamy ecology of the Alabama Black Belt, using the farmhouse as a springboard for childhood adventure. The youngsters spent most afternoons escaping the heat by swimming in the creeks and ponds. They would ride the running boards of the old farm truck the two miles down to the beaver dams where the water was best. After beating the surface with a paddle to scare off any reptiles, the children would scream and splash their way through the hot afternoon. On the drive home, the running boards provided a great place both to dry off and to survey the farmland as the truck bounced across the fields toward home. Back at the farmhouse, Rolston remembers the sweet taste of watermelon, cut fresh each day for a midafternoon snack.

Fishing was a serious pastime for the children in Alabama. The railroad trestles over the creeks made for some productive holes. The youngsters left baited trotlines out overnight to catch gar, shiners, eels, and suckers from the swamp water. Rolston and his cousin Bill Forbes also spent many exciting nighttime hours frog gigging. The two boys would launch a small rowboat and pan a searchlight across the surface of the water until a pair of eyes flashed back at them. While one cousin kept the light on the frog's eyes, the other would paddle carefully toward the mesmerized amphibian until they were close enough to gig it. Not yet conservationists, Rolston and his cousin cut off the frog's legs and threw the rest of it away. Their grandmother would fry the legs with butter and chives for the table the following day.

Rolston later wondered how close he came to getting himself killed on some of these nighttime escapades. While there were no alligators at the time in Bogue Chitto, cottonmouth water moccasins moved quietly above them in the branches of the oaks. Cousin Bill, later to become a farmer on that same land himself, remembers on one occasion quietly closing in on a frog with the gig, only to suddenly scream out in the night for his cousin to backpaddle away from a branch on which a cottonmouth lurked. In his panic, Rolston crashed them into another tree nearby on whose branches hung three more.

The children were given various tasks around the farm. Washbasins needed to be filled and porches had to be swept. There was a large kitchen garden by the farmhouse and the family ate nearly all of their food straight off the land upon which they lived. Their grandfather taught Rolston and his cousins how to milk. Water pumped from the cistern kept the milk containers cool during the day. If the children preferred their milk warm, they could drink it straight from the cow. Rolston was never a very good milker, though he stuck with it and got better. Summertime on the farm was not quite self-sufficient but it was close. "All you need from town is coffee and salt," his grandfather used to tell them.

Philosopher Alasdair MacIntyre, convinced that ethics requires a story line from which to emerge, has drawn attention to the important role of family in shaping the initial contours of one's moral point of view:

> I am someone's son or daughter . . . someone else's cousin or uncle. . . . As such I inherit from the past of my family, my city, my tribe, my nation, a variety of debts, inheritances, rightful expectations and obligations. These constitute the given of my life, my moral starting point.

Rolston's family, like many others in the South, always attached great significance to its ancestry. Later in life, Rolston would feel an increasing influence of the first Holmes Rolston, his paternal grandfather, who had died eight years before he was born. During his early years, however, Rolston gained a large part of his moral starting point from Will Long, the grandfather with whom he spent the summers exploring the hot dirt surrounding Bogue Chitto Creek.

On late summer evenings, with the moon beginning to rise over the slash pines, Rolston and his cousins would sit on the porch of the Alabama farmhouse swatting bugs and watching intently as their grandfather settled himself down to tell a family story or two. On these evenings, Rolston learned from Will Long some of the cherished details of his family's rootedness in the southern landscapes.

Long was the product of a line of farmers and adventurers. His own grandfather, Daniel Long, had followed a winding and occasionally hair-raising path to Alabama. Daniel Long had planned to be a medical doctor and was in training in Philadelphia when the war of 1812 erupted. Called into service as a medical assistant, he was dispatched to the gulf coast, where he tended the injured and dying as best he could with his incomplete skills. Returning home at war's end without enough money to continue his studies, Daniel Long decided to see the country. He got on a horse in Philadelphia and rode it to the gulf coast of Texas, sleeping in hayricks and woodlots along the way. Like John Muir on his walk to the gulf half a century later, Daniel Long used the solace of an extended journey through the South to settle his mind, finding nourishment in the long days of cross-country riding and quiet nights gazing far into the inky blackness.

After two years working in Texas and saving his money, Long returned to Philadelphia to complete his medical studies. On graduation he found some land in the territory soon to become Alabama and moved into a large house where he ran both a medical practice and a cotton farm.

Like most cotton growers in mid-nineteenth-century Alabama, Long owned a number of slaves. Both a doctor and a business man, he saw the importance of keeping those who worked for him healthy. Whether he did it out of a physician's compassion, a calculated business mind, or a sense of Christian brotherhood is not known, but Long gained a reputation among his neighbors for the care and attention he gave to his slaves. It was care that also brought him a share of personal tragedy.

Called out one night to treat a slave child who had contracted whooping cough, he took the gravely ill girl into town in his carriage, covering her with his coat. After a late night return home, Dr. Long was out again early the next morning to attend to some business on the farm. When he got back to the house later that day, he found his daughter playing in the same coat he had used to take the slave girl into town. A few days later his daughter also came down with the disease. She died painfully within three weeks.

Daniel Long's grandson, Rolston's grandfather, was born during the difficult years of Reconstruction. Will Long spent most of his childhood working alongside his parents to keep the farm from going bankrupt. He married Willie Francis Smith in the 1880s and they made a start with ten acres of their own near Uniontown, Alabama. Fifteen years later he brought the farm three miles west of Marion where Rolston spent his summers. Going against all economic sense, Long insisted on keeping the new farm partly in hardwood forest. Though it was always a questionable decision, he liked the idea that he was now comfortably enough off to keep some of his land in trees rather than work every square inch for market. Besides, Rolston's grandfather loved the woods and creeks.

Long possessed the same attachment to horses that his grandfather Daniel Long had possessed before him. Horsemanship had lodged somewhere in the family blood. A legendary great-aunt Susan had been killed at the age of ninety when jumping her horse on a morning ride. Will Long always dreamed of owning enough land to spend a whole day on horseback and still not get around his property. He never managed this, but Rolston has memories of his grandfather frequently riding off by himself over his fourteen hundred acres to check for the umpteenth time that everything was still in order.

It was from Will Long that Rolston picked up the first fragments of an environmental ethic. In the absence of formal schooling, Long was taught to read by his mother who had attended a "finishing school" in antebellum years. Though the Alabama swamps bore little resemblance to the Lake District of northwestern England, Long gained from his mother an interest in the English romantic poets. Rolston remembers his grandfather quoting Wordsworth and Shelley on the farmhouse porch. Long lived by a kind of primitive environmental ethic, cautioning his grandchildren to "Take care of the land, and it will take care of you." Unlike many rural Americans of his age, the Black Belt farmer recognized that the earth was not to be squandered. "Be careful with your land," he would warn. "They are not making any more of it." He knew that the fields represented more than just a source of income. Deeper moral and religious values appeared to lie barely concealed within the earthen clods. Will Long passed an early appreciation of these values on to his grandson.

Rolston's grandfather continued to ride his horse around the farm until three months before his death at the age of eighty-six. When his joints were too arthritic to mount his horse from the ground, he led her into a ditch near the water pump and circled back to climb on from the bank. By this time, his wife had begun to fear that her husband would suffer the same fate as Great-Aunt Susan. A well-behaved southern lady, Willie Francis did not dare tell her husband not to ride. But she knew how to achieve her goal. One evening she made sure her husband's saddle was left unattended in the barn overnight, in full view of the farmworkers. As Mrs. Long had hoped, the saddle was stolen before dawn and Will Long never rode his horse again.

When Rolston recalls the impression Will Long left on him, he speaks with plenty of admiration for his hardworking grandfather's character. Alabaman agriculture in the 1930s required more in the way of hard graft and thrift than technology and complicated machinery. Farmers and their "hands" still plowed with mules, hoed the corn, and picked cotton manually. Rolston has lingering memories of his grandfather rooting around the barn for scraps of metal, then firing up the shop forge to fix a broken carriage or plow. With one of the black farmhands pumping the massive bellows, Long would fashion the glowing scraps into plow blades or singletrees for the wagons the next day. "Willful waste makes woeful want" was his grandfather's adage, and waste was something this southern farmer could ill afford.

The summers on the Alabama farm impressed upon Rolston the idea of a primitive and essential goodness located deep within human nature. One of the ways this nature expressed itself best was when humans worked the land, something Rolston recognized again years later when he traveled to Africa and rural Asia. During one of these overseas trips nearly fifty years later, he wrote in his trail log: "There is integrity written into the lines of their weathered faces. They are poor in material goods, but the timeless virtues are here—love, honesty, courage, resolution, endurance, thrift, hope and laughter." He wrote these lines about some Nepalese villagers but he could just as easily have written them about Will Long and other farmers he knew from his youth.

After his grandfather's death in 1951, Rolston kept Long's blacksmith anvil to remind him of the "muscle and blood" economy his ancestors had relied upon to carve out their place in the world. Rolston also treasured an old butcher's knife from the farm on Bogue Chitto Creek, worn thin by repeated sharpening over half a century.

Virginia and Alabama helped sculpt Rolston's mind during these formative years. Each piece of landscape he explored provided more for the boy's senses to experience, more to be challenged by, and more for his mind to wrap itself around. The southern Appalachians became a part of him, supplying some of the structure that guided his later thoughts about the world. A rural Virginia manse with its vegetable patch and nearby river, its songbirds and amphibians, and an Alabama farm with its cows and cottonmouths, its pea vines and potato plants—all found their way into his brain, coloring the way that he saw his fit among the human and natural communities into which he had been born.

It is probably too much to suggest that by the age of eight a child has developed any kind of philosophical orientation to the world. Yet faint traces of Rolston's later thinking had already established themselves within his capable young mind. There were powers lying deep within nature and an order to how it all worked. The cycles of life and death and the unremitting contests for survival he witnessed in both the Shenandoah and in Alabama had imprinted themselves on his brain. The glorious transitions of color that emblazoned the Appalachian fall and the aggressive rebounding of life in the buds and blossoms of spring provided a compelling rhythm to his observations. The annual pilgrimages to Alabama helped give him an appreciation for how life and death in nature followed perennial patterns. Rolston also picked up from his elders the sense that there was something of moral, or even religious, significance lurking within the landscape.

At this point Rolston was still just a boy, little different from the tens of thousands of other children being raised in Depression-era America. Ahead of him lay a different kind of education into the theories, facts, and cultural environments of school and college life. Here, as we shall see, Rolston's development starts to take a more revealing form.

2

Country Schoolhouse to Atomic Physics Lab

> Unlike coyotes or bats, humans are not just what they are by nature. Humans come into the world by nature quite unfinished and become what they become by culture. Being human is more than biochemistry, physiology, or ecology. Humans superimpose cultures on the wild nature out of which they once emerged. [*1998*]

SINCE HIS NOVEMBER BIRTH DATE made their eldest child too young to enter the school system in 1937, Holmes and Mary Long Rolston homeschooled him in first grade. With his father keeping a busy schedule at the church, Rolston's mother took most of the responsibility for the academic duties. She appreciated the chance to be with her firstborn a little longer and to instill in him more of her Presbyterian values.

At this age, faith lent only a pale shade to the boy's thoughts. Pastors' sons are no different from other young boys who have important fishing trips to attend to and forest explorations to complete. Church was a boring imposition on the weekend schedule. On one sulking and tearful Sunday morning walk from the manse over to the morning service, Reverend Rolston tried to lift his son's spirits by pointing to a bird singing loudly in a nearby bush. "Listen,

Holmes! There's a bird singing, the one sitting there high on that branch. Isn't he pretty?" Without lifting his eyes from the gravel underfoot, the pastor's young son replied with the faultless logic of childhood, "The little bird is singing because it does not have to go to Sunday school." Rolston clearly did not feel his vocation early.

Despite the predictable frustration, Rolston still absorbed from his parents and his Sunday school lessons an abiding sense that he lived within God's beautiful creation. The fullness of the earth, his parents told him, was a divine gift of the most wonderful kind. The Lord had created the world in a particular way for a particular purpose. For this his people should thank and praise him. Attending church on a Sunday, his parents told him, was the least that a good Christian could do to express some gratitude.

The young boy had to admit that the fields and streams where he spent most of his afternoons and evenings did have the quality of a gift. There was so much of fascination to be found and so much impressive detail to explore. Virginians did seem to be the beneficiaries of something remarkable. This early message about God's providence struck a lasting chord.

After this first year at home, Rolston began classes in a three-room schoolhouse at Rockbridge Baths. The school was a simple wooden building without water or power, attended by some three dozen local children. Rolston took his lessons from three valley teachers. One of them was Clyde Tolley, an elder in his father's church. Tolley was both a teacher and a farmer, living in a log cabin on the road between Bethesda Presbyterian and the town of Lexington. Rolston has fond memories of Tolley, calling him "one of the finest men I ever knew."

Rolston took classes happily at the rural schoolhouse for several years, learning how to read and write and getting his first exposure to mathematics, science, history, and geography. At recess, the river was only a few dozen yards away and the children often brought some frog or bug back to the schoolroom for everyone to see. Rolston was an industrious student, full of questions and curiosity. The hours passed quickly. Under the tutelage of Tolley, Rolston began to learn formally something he had started to pick up on his own in the

Alabama and Virginia countryside. There was an order in the world to be learned. His lessons in the three-roomed schoolhouse taught him that the study of science was the key to unlocking that order.

When the teacher rang the bell for the end of the school day, Rolston's classroom again became the woods and fields of the Shenandoah. He delighted in being able to fish on Hays Creek or swim in the Maury after a full day at school. Back at home in the evening, his father quizzed him about what he had learned that day. Both parents followed his progress in school closely, making sure that the learning environment in the Bethesda manse was continually stimulating. Rolston's intellectual appetite sharpened by the day. Unfortunately, these bucolic years of early education were about to be interrupted. Shortly after Rolston's ninth birthday, his father came home from a parishioner's house one evening looking uncharacteristically worried.

It was early Advent in December of 1941. Christmas trees sparkled in many valley living rooms. The pastor had been called out to pray with a family reeling from a recent death. To the children sitting around the tree back home it seemed as if their father was gone for an eternity. The family rule was that when father was away, nobody should light the candles on the tree. Naked flames were too dangerous. The children fidgeted impatiently, begging their mother to bend the rule. Mrs. Rolston was resolute in waiting for her husband's return. When he did finally return, the children pounced on him as he came through the door, begging him to go immediately to the family room to light the candles. The pastor managed to raise some excitement for the benefit of his children but his wife knew right away that something was wrong.

The family sat and watched the candles burn to their bases. When the last wisps of smoke had disappeared, Mrs. Rolston turned to her husband and asked in a compassionate voice, "What's wrong, Holmes? You don't look happy. You are concerned about something." The pastor looked at the tree for a few seconds more. With carefully chosen words, he started telling his wife and his three small children the news.

While he was out there had been an urgent broadcast on the radio. The Japanese had bombed Pearl Harbor in Hawaii. More than a thousand Americans had died and several ships had been sunk. The nation was set to declare war against the Axis powers. America would join the carnage taking place far away on the other side of the Atlantic. The radio announcement and the errand from which the pastor had just returned formed an ominous pair. Rolston's father could anticipate the months and years of suffering into which the country would be plunged.

In a coincidence that did nothing to calm the family's unease, Sunday, December 7, 1941 was the day Holmes Rolston II had already made one life-changing announcement. A couple of hours before the Japanese planes had launched their surprise attack on Pearl Harbor, the pastor had informed his congregation that he would be leaving Bethesda to take up a position at West Avenue Presbyterian Church in Charlotte, North Carolina. His church superiors wanted to move him to this troubled urban parish in need of a pastor with experience. Rolston's father felt obligated to accept. It meant that he and his family would soon be leaving the Shenandoah Valley for a radically different world.

This was a difficult time for the children to move from Rockbridge Baths. Rolston remembered being distraught at the prospect of leaving the creeks and woodlands where he felt so much at home. He had also picked up on his parents' anxiety about the move to urban North Carolina. He informed a relative in a letter at the time that his feeling was one of "disgust." His father tried to cheer him up by telling him there would be the opportunity to see airplanes in Charlotte. Their new home was said to be near the airport.

Although he sulked openly about the move, deep inside a hidden part of him was curious and a little excited. On his next birthday, Holmes would be ten. He had never lived outside a rural valley. He recognized the possibility for new adventures. No doubt he would meet plenty of children in the city as well as see airplanes. Much of what he had learned in Clyde Tolley's schoolhouse suggested that the important things in the world happened in big cities. On the day they drove out of the Shenandoah Valley and pointed the car toward North Carolina's Piedmont, Rolston's stomach churned with a strange mixture of sorrow, dread, and excited anticipation. The last thing his father had done before leaving the manse was to shoot the family cat. There would be no space, father had announced, for a country cat in the city.

Rolston's memories of his time in Charlotte turned out not to be happy ones. He never saw many airplanes. The house they had been assigned was farther from the airport than promised. He did make a few friends but that benefit paled beneath the grim atmosphere of the war. Many of the city's men under the age of forty were drafted to fight in the European and Pacific theaters. Large numbers did not return. His father's role as a spiritual counselor placed the family squarely at the center of the grief that was a perpetual presence in Charlotte. Those who escaped from suffering personal loss still felt the pervasive anxiety caused by rationing and the constant stream of depressing news from the front.

One of the few times Rolston felt any direct contact with the war was when his father took him down to the rail yard to look at a trainload of German prisoners being shipped to a prison camp. He remembered being surprised at how normal the Germans looked. They were not as mean-looking as he had expected Nazis to look. At the same train station a couple of years later, he saw a triumphant General Dwight Eisenhower, commanding officer of the victorious forces and a future president, waving to an ecstatic crowd of wives and demobilized soldiers from the back of a VIP train. Rolston felt a child's puzzlement about how people who looked so alike could find themselves wanting to kill each other.

The war was not the only dark cloud that Rolston recalled from his Charlotte years. No sooner had the war ended than the city began to suffer again. The late 1940s marked the beginning of a polio epidemic in the United States. During several of the summers in Charlotte, the city authorities quarantined all children in their yards in order to minimize their exposure to the virus. A number of Rolston's school friends contracted the disease and were crippled for life. A few of them died. Others ended up in iron lungs from the paralyzing effects of the disease.

Rolston also struggled for a long time in Charlotte to make the transition from a rural to an urban child. Unable to wander barefoot across fields and forests that had long ago disappeared under brick and cement, he looked for new pastimes. He played chess with the children next door, half the chessboard resting in each family's yard. The trains at the nearby rail yard and the smokestacks from the local power plant pumped large quantities of soot into the air. Rolston remembers the white squares on the chessboard filling up with particles that dirtied the children's fingers and turned their handkerchiefs black. When not quarantined against polio, he made pocket money by pushing mowers around the neighbors' lawns, using his youthful energy to advantage in the oppressive heat and putting the money he received into government war bonds. As Rolston sweltered in Charlotte's summer humidity, his mind went back to the relative cool of the Shenandoah Valley, his memories of the Maury River acting as a salve for those long summer days and as a lure toward better times ahead.

During his eight years in Charlotte, Rolston was enrolled first at Vance Junior High and then at Harding High School. Compared to Clyde Tolley's three-room schoolhouse near Rockbridge Baths, both of these schools were vast and intimidating. Harding was filled with tough kids from working-class neighborhoods. Rolston was initially known to his classmates as "PK," the preacher's kid. A number of the students made fun of him for his country accent.

The teasing did not last long. By this time, PK was beginning to shine academically. The schoolyard bullies realized they were dealing with no simple-minded country boy and turned their attention elsewhere. Rolston's grades earned him recognition in the *Charlotte News* for being one of only four students in a class of fourteen hundred to maintain straight As through high school. He was elected a member of the National Honor Society. He was particularly good at mathematics and the natural sciences. A few of his classmates called him "egghead" but most of them envied his smarts. A young girl named Patsy remembers him as "tall, gangling, cussing PK—smart as they come." When PK walked into the room, Patsy blushed.

As his knowledge grew, Rolston's excitement for the natural sciences started to build, fueling the curiosity he had first felt roaming through the Shenandoah Valley and wading in Bogue Chitto Creek. He found himself enthralled by the power of numbers to read the world. He began to see how math and science made it possible to manipulate the world to human advantage. His father bought him a better slide rule than any other of his classmates and Rolston obsessed over it, mastering techniques that most of the other children in school did not even try. He thought that he might become an engineer or an inventor.

In 1947 his parents gave him a chemistry set for Christmas. He set up a laboratory in his bedroom and explored the alchemy he could create with test tubes and pungent, colorful liquids. He bought himself a telescope and a cheap camera with money saved from working in a downtown store and experimented with developing his own photographs. Like many other youngsters in the 1940s, he built radio sets, prowling the garbage cans behind the radio repair shops to find discarded parts. He fashioned a device using an old alarm clock for switching on the radio in the morning. Through magazines he learned how to insulate the end of the clothesline outside in the yard and use it as an antenna. He ran a wire from the line to the upstairs porch where he slept in the summer heat. There at night he secretly listened through headphones to the shortwave radio he had built, his mind reaching far beyond his Charlotte home to distant and exciting worlds.

Toward the end of his time in high school, Rolston was appalled to learn from his father that the family would be moving again. This time they were marching north to the town of Richmond, capital of their native Virginia. His father had been offered a position as an executive editor at the Board of Christian Education of the Presbyterian Church. With several books published in Protestant theology, Holmes Rolston II had slowly been gaining a regional reputation as a theological scholar. He maintained an intellectual correspondence with the famed German theologian Emil Brunner and had hosted Karl Barth's visit to the Civil War battlefields during the Swiss Calvinist's only trip to the United States in the 1950s.

It was not, however, the Charlotte pastor's connections with Brunner or Barth that had landed him the invitation to work for the board. Rolston's father had for years used the time his children were down in Alabama to become an accomplished Christian educational writer. He transposed his sermons and published them in church quarterlies as "Adult Uniform Lessons." His insights into the meaning of biblical passages and his ability to talk about them in everyday language quickly won him an enthusiastic readership. Soon the lessons were being distributed through regional quarterlies where they reached about 75,000 people each week. The invitation to serve on the board of Christian education would allow Rolston's father to draw on these skills to help shape the academic curriculum at Presbyterian schools throughout the region. Making a career switch that would be repeated twenty years later by his son, Rolston's father resigned from his work as a pastor and readied himself for a new career in education.

The family moved to Richmond in 1949. For Rolston, this meant starting a different high school in his senior year. Just as it had in the Shenandoah Valley seven years earlier, his slowly emerging social life felt like it was being crushed under the wheels of a car filled with family belongings. If the young man was not already inclined to put academics before friendships, the timing of his father's moves was forcing him to do so. The teenage Rolston started to become more and more introverted and increasingly interested in books and learning.

In the end, there was not much time for Rolston to settle into Richmond. With the end of high school looming, it was already time to make decisions about college. After being in town for only nine months, he graduated from Thomas Jefferson High School in Richmond as a straight A student in the summer of 1950. The following fall he traveled back over the border into North Carolina to enroll at Davidson College, one of the South's most well-known liberal arts colleges located just outside of Charlotte.

Though his stay in Richmond had been brief, there was at least one life-changing consequence of his year in the Virginia capital. It was in Richmond that he first became acquainted with a local girl named Jane Wilson. The two had met on Rolston's first Sunday in town at Ginter Park Presbyterian Church. When he introduced himself to the Sunday school class, Jane remembers Rolston saying rather awkwardly, "It's Holmes, as in Sherlock; Rolston, as in Shredded," using the name of the detective and a popular breakfast cereal to illuminate the pronunciation. He was a slender and intelligent-looking young man. Jane was immediately curious. Later in the day the two were introduced to each other directly by a girl Rolston knew from church retreats in North Carolina. His first impression was that Jane was "an exceptionally pretty young woman." Though they hardly spoke that first Sunday, the interest of both high schoolers had been sparked. For a year they saw each other weekly at the youth activities at Ginter Park. Days before he left town for Davidson, Rolston took Jane to a radio show for their first date. Before either of them had time to think about what might come next, Rolston was on a Trailways bus heading down the highway in the direction of Charlotte.

Davidson College was a fortunate choice for the development of Rolston's growing intellect. The highly regarded Christian institution provided exactly the right mix of religious and secular learning for the talented undergraduate. Founded by Presbyterians in 1837 on four hundred and fifty rural acres, the college had as its motto *Allenda lux ubi orta libertas* (Let learning be cherished where liberty has arisen) and claimed to celebrate both Presbyterian and American values. The college's deep roots in Calvinism blended with the faculty's genuine interest in encouraging students to be tolerant of other religious traditions. The mission statement made it clear that the loyalty of the college extended "beyond the Christian community to the whole of humanity and necessarily includes openness to and respect for the world's various religious traditions."

More important for Rolston at the time than its openness to different religions was the openness Davidson also showed toward the natural sciences. The college recognized the importance of both science and religion in understanding one's place in the world. Professors encouraged in their students the idea that "faith and reason work together in mutual respect and benefit toward growth in learning, understanding, and wisdom." This willingness to search for a respectful relationship between these perennial antagonists left a lasting impression on the young Virginian.

Rolston came to Davidson determined to explore how numbers could unlock the secrets of the universe. He quickly chose to major in physics and mathematics, studying under motivated professors in the grip of the excitement of the new atomic age. His talent with numbers soon made him a calculus tutor to his fellow students. Week by week, Rolston was drawn deeper into the mathematical mysteries of nature, equally compelled by the workings of the atom and by questions about the origin of the universe. It did not matter if his tools were cloud chambers, telescopes, or microscopes. Traveling outside of humans' native perceptual range promised access to the deep truths he now fervently pursued.

Reminiscing later about this period of intellectual growth, Rolston wrote,

> I entered Davidson College in 1950 eager to learn the mysteries of physics. [Physics] seemed to be the science of fundamental nature and I was . . . attracted by the physicist-philosophers probing nature in the very small and very large, microscopes and astrophysics, and the cosmology that results from philosophizing over discoveries at these ranges. . . . Perhaps there was nothing to learn from trees and rustic places, but there was everything to learn about matter-energy from cyclotrons and Geiger counters in town. This wasn't wild nature; it was mathematical nature. At the bottom of it all, there was ordered harmony, symmetry, universal law, beauty, elegance.

This most ultimate science gave Rolston all the intellectual stimulation he needed. What he did not yet realize was that it would soon lead him to a rude collision with his religious beliefs. At the time, however, the only worry he had about his studies was more practical.

Rolston missed his explorations of the Shenandoah Valley and the close contact with nature they provided. At Davidson he was constantly studying how nature worked in some ultimate theoretical sense, but he did not feel much of a connection to nature in any everyday sense. Craving a break from the mathematical detachment, Rolston got out as much as he could into the local woods with his Davidson roommate, Aubrey Miree. Miree was a hunter and a crack shot who broke numerous records on the college rifle team. Both of them were nostalgic for the rural and mountainous backwoods of Appalachia. The afternoons Rolston and Miree spent combing the wild country of the Carolina Piedmont were always rejuvenating. Enjoyable as the woods and streams might be for recreation, however, outings there were starting to feel increasingly insignificant compared to the deep truths of quantum physics. Modern physics represented a different arena of discovery and Rolston was becoming firmly ensconced in that new framework. He probably would have abandoned the woods entirely had it not been for the influence of one particular and slightly eccentric biology professor.

As part of his science curriculum, Rolston was required to take a few classes in biology. Even though physics now had intellectual priority in his life, he applied himself to his studies in biology with customary diligence. His teacher, Tom Daggy, immediately created an impression on him. Daggy approached his field like a national research scientist rather than a teacher at a small Carolina college. To a young man with Rolston's intellectual drive, Daggy was a spectacular role model. The study of science seemed like a life and death matter to the young professor, one that justified spending evenings and weekends dissecting insects, tracking migrations, and chasing down errant biological facts. Rolston never had much of a problem getting motivated academically. Hanging around Tom Daggy, he learned that other interests came a distant second when there was some serious insect gathering or plant taxonomy to be done.

Rolston gave up spring break two years in a row to travel with Daggy and a small group of other motivated students to southern Florida to do hands-on natural history. "Buggy" Daggy, as the students called him, led the eager students through knee-deep swamps and insect-infested mangrove thickets into the heart of Florida's subtropical world. Daggy introduced them to the abundant marine life of the Florida Keys and to the exotic plant and animal communities of the Everglades. The students followed dutifully behind their professor, scratching notes onto damp sheets of paper, and peering through hand lenses at flower heads and beetles. Glass-bottomed buckets allowed them to glimpse the breathtaking marine world beneath the surface of the water. There they encountered coral reefs peppered with sponges, anemones, urchins, and sea cucumbers. Spider crabs and lobsters skittered feverishly across the sandy bottoms. Above the water they marveled at egrets, herons, and roseated spoonbills roosting in mangrove and cypress trees. Gallinules, coots, and anhingas paddled comically along the surfaces of the ponds, while alligators basked lazily on sun-drenched mud banks.

The students were mesmerized. Daggy could somehow draw their attention to things others were not seeing. When sitting around an evening campfire swatting the bugs drawn to the firelight, Daggy kept his eyes peeled for anything interesting. When he spotted an unusual insect, he would jump up and yell "Stop! Don't swat that one!" and use his tweezers to pop it into vial of formaldehyde he carried in a shirt pocket, informing his students, "We will look at that more closely when we get home."

Rolston learned that fieldwork was hard and required spending numerous hours laboring in uncomfortable conditions, with headaches and a generous spackling of bug bites for his troubles. Despite the discomfort, the young Presbyterian student gained a wealth of firsthand knowledge about the biodiversity supported by the tropical latitudes. He marveled again at the profusion of life and wondered afresh at the phenomenal richness of God's creation.

Months after one of these trips to Florida, Rolston was working late in Daggy's lab at Davidson completing a report. Suddenly Daggy called Rolston over to look at one of the insects they had brought back from the Everglades. Feeling puffed up that his professor had called on him, Rolston sauntered over, removed his glasses, and peered down the binocular microscope at an insect vaguely familiar to him from one of the sweaty excursions earlier in the year. When his eyes had the time to focus and his mind to absorb enough of what he was looking at, Daggy leaned over and said in a voice scarcely concealing the excitement it contained, "That, Rolston, is a creature new to science."

It was a pivotal moment for the pastor's son, one that pushed back open the door that his studies in physics and mathematics had been slowly closing. In that instant, Rolston grasped that there were important secrets of the universe right under his nose that did not have to be unlocked by physicists and astronomers. Biology need not be overshadowed or made irrelevant by physics. It too could reach into the heart of things.

In a moment of insight that was to guide much of the rest of his intellectual life, Rolston was struck by how the very presence of life was probably the single most startling fact in the universe. The operation of the universe according to mathematical principles was certainly fascinating. But the emergence of life on one of its planets seemed more remarkable still. Earth, after all, could have been just a lifeless hunk of rock. Instead, it was home to millions of colorful species interacting in complicated and marvelous ways, not just sustaining themselves, but also multiplying and diversifying. Now *that,* thought the young scientist, was something worth investigating. Wading around the Everglades with Tom Daggy made him think that perhaps it was biology and not physics that cut to the essence of things. Physics and mathematics could only ever tell him about linear order and structure. Biology could tell him about life.

As he lay in his bed with his eyes wide open on the night of the encounter with the new insect, his mind flew between the oceans and the stars. He was energized by the intellectual voyage he now saw lying ahead. He could already see huge puzzles needing explanation, ranging from questions about the birth and origins of life to mysteries about speciation and extinction. Why was earth like this? What gave these biological processes their direction? Where does one go to find theories to explain these mysteries?

Rolston was also smart enough to sense the clouds beginning to gather on the horizon. Questions about the origins of life clearly had theological dimensions to them. Religious and nonreligious people were bound to approach matters of origins differently. His teachers at Davidson acknowledged the challenges presented by evolutionary theory. When Rolston asked them about the tension, they suggested that modern biology could be readily reconciled with the Christian faith. It just had to be done carefully. But Rolston already noticed the lack of detail in their replies. None of his teachers ever offered much in the way of concrete ideas about how to achieve this reconciliation. With graduation from Davidson drawing closer, his investigations of these big questions started to mix with thoughts about fundamental matters of a quite different kind.

The whole time Rolston had been at Davidson, he had never been able to get Jane Wilson entirely out of his mind. Serious student though he was, he found his imagination often wandering north up the fault line to Richmond and the girl he knew was attending Ginter Park Presbyterian Church on Sunday mornings. He sent Jane a postcard from the first of his Everglades trips and signed it "Me." It took Jane weeks to figure out who it was from. As he progressed through college, his thoughts of the young woman he had met in Richmond seemed to occupy more and more of his time.

Jane was by this time studying classical languages at Westhampton College, a women's division of the University of Richmond. They saw each other during Christmas and summer breaks. Jane and a friend rode down to Davidson on the bus to attend the fall dance in 1951. With increasing frequency, Rolston stole away from classes and hitchhiked the two hundred and eighty miles to Richmond to see her. He would pick Jane up in his father's prewar gray Pontiac and they would head to the ice cream shops and city parks.

These visits were not without their dramas. During one of the illicit trips, a Richmond police officer gave Rolston a ticket. Rolston's father had a journalist friend who happened to find out about it. The journalist was at the police station checking up on some other story when he ran into the officer who had stopped his friend's son.

The journalist asked the cop what exactly had happened in Rolston's brush with the law. The officer took off his cap and chuckled:

> I pulled him over for not coming to a complete standstill at a stop sign. Soon as I walked up to the car I felt so sorry for the young kid. I hated to give him that ticket. He was sitting next to the prettiest young lady I ever saw. I wouldn't have noticed the stop sign myself if I had that young woman sitting next to me.

At this point, Rolston felt as if he was doing most of the work persuading Jane of the merits of the romance. Though he was more accustomed to spending time with books than with members of the opposite sex, he had clearly decided his labors were worth the cost. Only a month before he was due to graduate, he cut classes to see Jane performing as maid of honor in the Westhampton College May court. His academic dean, John Bailey, wrote him up with a warning that he would not get his degree if he cut any more classes. He was grounded until graduation.

Rolston was now within weeks of leaving Davidson with a double major in mathematics and physics. He had been elected a member of the honorary scholastic society Phi Beta Kappa and was clearly an exceptional student. In the article in the college paper announcing his advancement to Phi Beta Kappa, Rolston described himself confidently as "a physics major who intends to enter graduate school, probably the University of Chicago, next year and who will shortly teach or do research in physics." He already had offers in hand from several graduate schools.

Beneath the confidence and polished exterior, however, something he had not yet shared with anyone but Jane was creating an increasing turmoil inside. Much as he enjoyed his work in the sciences, his thoughts had recently begun to turn in an entirely different direction.

3

Graduate School to Rural Pastorate

> Some accounts find religion to be less linguistic and thus less logical than science. This may be taken by critics as a vice, but it also may be taken by proponents as a virtue, that religion plunges to deeper levels than the conventional ones of science. This latter position is not without merit, for the religious object, God, if it exists, is incomparably greater than any routine scientific object, such as rocks, fish, or atoms. [*1987*]

TO HIS SCIENCE TEACHERS at Davidson, it came as a bit of a surprise to learn that one of their star students had decided to go to seminary. Rolston had shown such a precocious talent for the natural sciences. The idea he might set all this cutting-edge learning aside in order to delve more deeply into theological questions was a disappointment to some of them. When he told his advisor at Davidson that he would not be attending graduate school in physics, the advisor replied wistfully, "I only wish there were two of you."

Within the larger contours of Rolston's life, the move from physics to theology was not the surprise it first seemed. Most students approach graduation with some nervousness about what is going to come next. Rolston's family history suggested the church might be a reasonable choice in the face of the uncertainty. Presbyterianism had played a large role in shaping the lives of Rolston's ancestors in the New World. In addition to his father and grandfather, a great-uncle named Calvin Rolston had chosen a religious life as a missionary to the Pawnee Indians in Oklahoma. Rolston's grandmother had once written to him in a letter, "Maybe some day you will follow in your Daddy's and your Grand Daddy's footsteps and become a minister. I couldn't ask anything higher for you." Though his father did not put any overt pressure on him, some of the elders in his father's church indicated not too subtly that seminary was a fine choice for an academically gifted young man. Rolston later confessed that in times of uncertainty like this, "sometimes you do what those you look up to tell you to do."

A clear intellectual trajectory was already taking shape in the young man's mind. Ever since staring at the hydraulic pump on the Maury as a child, Rolston had been intrigued by the thought of hidden powers lying deep within nature. When his scientific interests at Davidson had turned from applied to theoretical science the questions started to shift subtly from *how* to *why*. Quantum physics and astronomy steered Rolston's curiosity beyond his practical interest in engineering nature to good purpose and on toward deeper philosophical inquiry.

Rolston had chosen physics to find out how the world was made, but he was starting to find that physics always left a small residue of questions unanswered. Physics was widely regarded as the ultimate science, but it did not seem to have all the ultimate answers. Drawn to the explanations offered by his faith, Rolston wondered if maybe the theologians knew something about origins that the physicists and astronomers did not. Insofar as Rolston's Davidson professors had thought his education was shaping him into a potential scientific researcher, the decision to go to seminary was a startling turn. Insofar as Rolston's quest was ultimately about reaching a coherent understanding of meanings and origins, seminary was nothing but the obvious next step.

Nevertheless, Rolston tortured himself for months about the decision to change his path. When he felt his mind was made up, he took a day out from school to travel up to Richmond to tell his father. As his father listened with sensitivity, Rolston sobbed almost uncontrollably as he explained why he thought he should turn away from a life in science toward deeper questions of meaning and the divine.

Rolston anticipated there would be serious dissonances between his past scientific education and his upcoming theological one. There was a wide gulf between the languages and the methods employed by the two to explain worldly phenomena. In his bones, Rolston sensed that there must ultimately be a profound convergence between them. What this convergence would look like, he did not know.

In his final term break before leaving Davidson, Rolston traveled north to look at potential graduate schools, among them Princeton Theological Seminary. As a rural boy with a passion for physics, he was told there were two things he simply had to do while in New Jersey. One was to see the rotolactor and the other was to see Albert Einstein.

The rotolactor was the latest device for bringing modern, mechanical efficiency to the milking of cows. About thirty cows climbed onto a giant turntable and started feeding on hay placed at the center. A technician attached milking cups to each cow's udders as they began to eat. As the rotolactor turned, the cows munched contentedly and the machine collected their milk. By the time the platform had rotated a full circle, the milking was complete and the cows had been fed. The technician detached the milkers, a door automatically opened, and the contented cows exited one by one. Remembering his own clumsy attempts at milking on Will Long's Alabama farm, Rolston gazed at the rotolactor with a mixture of admiration and amusement. The muscle and blood economy he remembered from his summers in Alabama was rather a long way from today's technology.

Far more impressive than the rotolactor to the aspiring seminarian was Albert Einstein. Rolston had been told that to see Einstein he should stand down the street from the physicist's home in Princeton at 10.30 a.m. At precisely that time, he was assured, the great scientist would emerge from his front door and walk to the Institute for Advanced Study a few blocks away. Rolston followed the instructions and Einstein emerged from his home exactly on cue and walked past him only three feet away. The twenty-year-old from rural Virginia was far too nervous to say hello. Like any serious student of physics, Rolston stood in awe of Einstein's intellect. Simply being in the presence of this Nobel Prize winner rendered him speechless. His first thought after the encounter was of just how much the famed scientist looked like all the pictures he had ever seen of Einstein, with his shock of white hair and shabby tweed jacket.

At the time, it was Einstein's achievements in relativity theory that impressed the young physics major. Later in life, the admiration would shift toward how Einstein had used the genius of reason to look beyond the human perspective and reach toward ultimate explanations. "The eternal mystery of the world," wrote Einstein in 1936, "is its comprehensibility. . . . The fact that it is comprehensible is a miracle." Einstein's achievements gave Rolston a lasting respect for the power wielded by the human mind. The brain was capable of exploring enormous territories, continually probing the hazy intersection of mystery and understanding. Rolston sensed the nobility and quality of this quest. For the rest of his life, Rolston wished he had possessed the courage to say hello to Einstein that day and to meet the physicist's eyes.

In 1953 Rolston began his studies for a divinity degree at Union Theological Seminary in Virginia. His choice of seminary was an indication of just how much he was influenced by family tradition. Rolston's father had attended Union Theological from 1927 to 1928. In 1898, when Union Theological was still part of Hampton-Sydney College, his father's father had also studied there after leaving the family farm. Rolston had never known his paternal grandfather, but through the many stories told about him, he was starting to feel increasingly close to the man who first bore the family name. The choice of seminary was in part a small nod to his grandfather's memory.

In addition to the family history, there were other advantages of going to seminary in Richmond. Deep down, Rolston was a southerner and he did not want to move to Princeton or New York. Union Theological in Richmond would keep him near to the Valley of Virginia where he grew up. He would be able to make trips back to the woods and creeks he had roamed as a child, something that seemed appropriate as he now turned his thoughts increasingly back toward origins. Another prominent consideration was that studying at Union Theological would keep him in Richmond for three more years. Richmond ensured a close proximity to Jane.

At seminary, Rolston was assigned a former U.S. Navy sailor named Bob Crumby as a roommate. There could not have been more of a contrast between the two. Crumby was a boisterous athlete, more inclined to spend his spare hours playing pickup ball on the sports field than revising his tenses in the classical languages. Rolston occupied the other end of the spectrum. Uninterested in athletics, he relaxed, as Bill Forbes remembered down in Alabama, by sitting under a tree reading a book. His athletic prowess was limited and Rolston's peers joked at his lack of dexterity. He moved awkwardly across campus, earning the nickname "sidewinder" for the meandering gait he adopted as he walked between classes. Crumby and Rolston were selected as roommates to balance each other out. The friendship lasted a lifetime.

Crumby remembers just how much his roommate stuck out intellectually. Rolston excelled at quizzes and exams, consistently scoring near the top of the class. Outside of the assigned coursework, Crumby could not fathom how Rolston also found the time to learn how to type, to read music, and to play piano. He recalled his roommate leaving campus several evenings a week one semester to take a water safety class in Richmond for reasons never revealed, though perhaps connected to his near drowning in Hayes Creek as an infant. Crumby was also on the receiving end of at least one memorable show of anger from his normally placid suite mate. Rolston severely reprimanded Crumby for throwing a paper cup out of the car window when they were cruising around Richmond one summer's evening. In the 1950s nobody really thought much about littering. Crumby was startled at how deeply his roommate cared about the local landscape.

Despite the switch from physics to theology, Rolston found his studies at Union Theological coming relatively easily. He worked hard and made quick progress through the program. Seminary continued to expand Rolston's mind in directions that would be important for his later philosophy. Certain patterns of thinking were starting to solidify in his brain, patterns that would reappear years later.

His graduate studies gave him a whole new appreciation for the relation between history and ideas. Though he fully understood the value of history at a personal and family level, cherishing the tales told about his ancestors and diligently collecting various family heirlooms and memorabilia, studying theology gave him a sense of the equal importance of history to intellectual life. This was not something he had appreciated at Davidson. Physics, and in particular the atomic physics ascendant in the 1950s, looked almost exclusively toward the future. Gazing into a cloud chamber, Rolston doubted there was anything for the past to teach him about the atom. It was similar with the other sciences, including Rolston's new passion, biology. Science was about breaking through to a brighter future. Like many other smart young scientists, Rolston had felt something close to contempt for matters of scientific history, unwilling to spend his time or energy on theories that had been proven false.

After Rolston started to study patristics, biblical history, and Christology, and when he began to look closely at the development of Christianity through the ages, he realized how nothing in the intellectual world assumed its present shape without reference to the past. New approaches almost always used the theory they replaced as a point of departure. This was as true in physics and biology as it was in theology and philosophy. Intellectual claims were never entirely detachable from their historical context. Hence, whether you were talking about people, about nature, or simply about ideas, it was ridiculous to dismiss history entirely.

Rolston also learned important lessons from his studies of Hebrew and Greek. Ideas tended to be understood differently depending upon the language in which they were articulated. This difference was heightened when the languages used different alphabets. His previous studies of physics had led him to believe that words mapped the world in a tight one-to-one relationship. An atom was an atom whether gazed at in North Carolina, the Netherlands, or North Korea. The study of languages gave him a new insight into how the same concept could be understood in different ways depending upon the language used to describe it. Words gained their meaning relative to a whole network of other words and ideas. The Hebrew word for "wind," for example, could also be translated as "breath" or "spirit" or "mystical power." Its meaning in Hebrew and in English could never be identical if the words used carried such different connotations and associations. The capacity of language to capture precisely the nature of things, an essential background assumption when Rolston was doing physics, no longer looked quite so certain.

With these new appreciations for the importance of history and linguistic context, Rolston started pondering some far-reaching puzzles about the different ways humans try to understand the world. Because of the importance of context and interpretation, Rolston for the first time appreciated the valuable role of struggle, argument, and debate in intellectual matters. The Greeks had called the process "dialectic." Intellectuals argued not just to find out who was right, argument also honed the critical ability of a thinker to see what he or she really believed. The cut and thrust of dialectic sharpened one's concepts and categories, making them more fit to survive in future disciplinary contests. "Iron sharpens iron, and one man sharpens another" (Proverbs 27:17) became for a time Rolston's favorite proverb at seminary. It suddenly seemed less surprising to him that the church was continuously arguing and splitting into different factions.

But the underlying tension here was obvious. Neither history, nor context, nor struggle had seemed even remotely relevant to the understanding of the world Albert Einstein had proposed. Sometimes one had to understand the world through lenses that were fixed. Other times it seemed that one's interpretation could be more varied. In some respects natural phenomenal seemed fixed but in others fluid; an irreconcilable tension built both into the structure of things and into the ways we understood them. This tension between fixity and fluidity was to haunt Rolston for years.

Seminary marked a new level of intellectual maturity for Rolston. He no longer just soaked up the material handed to him by his teachers as he had done at high school and at Davidson. He was learning to think independently by abstracting out certain themes. In essence, he was becoming more philosophical. It was complex but fascinating stuff. With his head beginning to spin from the depth of the puzzles, Rolston was glad to find other parts of his life coming to rest on more settled ground.

The efforts Rolston had put into courting Jane Wilson while he was at Davidson were finally bearing fruit. Jane had completed her studies at Westhampton College just a week after Rolston had graduated from Davidson. Rolston rushed up to Richmond to attend her graduation ceremony. When he began seminary, Jane enrolled at the nearby Presbyterian School of Christian Education for her own master's degree. Every evening at 9:15 p.m. Rolston would set aside his work and walk down Seminary Avenue to the Wilsons' house to see Jane. In the middle of his first year at seminary, Rolston gave Jane his Phi Gamma Delta pin. Giving a fraternity pin to a date was like an "engaged-to-be-engaged" commitment, as Jane put it. When summer school and a voyage to Amsterdam as a community ambassador kept them apart, Holmes wrote Jane a letter every day. In the winter of 1955, Rolston had the fraternity pin remade into an engagement ring and they started making plans for a wedding.

Holmes and Jane set their wedding for June 1, 1956, at Grace Covenant Presbyterian Church in Richmond. Jane had always loved magnolias and was determined to carry one in the ceremony. A family friend watched her garden all through May for a magnolia to bloom. On the last Friday before the wedding, they drove out to the friend's house where Holmes climbed a ladder and picked several of the pink and white blossoms.

The order of the ceremony was taken from the wedding service Rolston's grandfather had used in his Valley of Virginia church at the turn of the century. Rolston sold the war bonds he had bought with the money made from cutting lawns in Charlotte in the 1940s and had a solid gold wedding band custom-made for Jane's finger. The newlyweds celebrated with a punch and cake reception in the gothic church parlor before leaving for a honeymoon at a cottage on the beach on the Florida gulf coast.

When they returned to Richmond after the honeymoon, they learned that the clergyman who had performed the marriage ceremony was not licensed to perform marriages in the state of Virginia. In the eyes of the law they were not yet legally married. The devout young Christians, just returned from their honeymoon week, were horrified. Rolston's father took the Reverend Robert Boyd aside and urged him to attain the necessary certification posthaste. Within three weeks, Dr. Boyd made good on his word and the Rolstons legally became husband and wife.

At this point in his life, now happily married and newly minted with a divinity degree from Union Theological, Rolston could probably have found a position in a rural Virginia church without too much difficulty. But he was not yet ready to settle down. There was one more piece of family history to honor before he took up a pastorate. Thirty years previously, Rolston's father had won the Thomas Cary Johnson Fellowship and traveled to Scotland to study at the University of Edinburgh. Scotland was hallowed ground for American Presbyterians. A comprehensive education in Presbyterian theology almost required a trip to its cold and blustery shores.

In his final semester at Union Theological, Rolston won the W. W. Moore Fellowship. This award was given to the student the faculty judged to have the most academic promise for further studies. Though tempted by the theology departments in Zurich and Basel, it was Edinburgh—the "Athens of the North"—that attracted him most. Rolston decided to follow in his father's footsteps and travel to Scotland for a doctorate in theology. There he would hear other footsteps echoing along the cold granite of Edinburgh's northern pavement. In addition to the Calvinist John Knox, there was the great philosopher and empiricist David Hume, and a certain influential biologist named Charles Robert Darwin.

The newlyweds departed New York aboard the liner *Queen Elizabeth* on September 26, 1956, bound for Southampton. They carried with them two trunks, one almost entirely taken up by the large mechanical typewriter Rolston planned to use for writing his doctoral thesis. As poor graduate students, they could not afford much in the way of accommodation. Their cabin was well below the waterline and, since they lacked the ten dollars needed to rent deck chairs, during daylight hours they mostly just walked around the decks. Throughout the five-day crossing, they frequented the ship's smoking room where, at an appointed hour, they listened to radio broadcasts from the BBC.

On arrival in Southampton, feeling slightly disoriented by the long journey and the unfamiliar environment, they took the boat-train to King's Cross Station, the main connecting point by rail for northern England and Scotland. King's Cross was an imposing and impressive piece of Victorian architecture, with cigarette smoke and soot from the locomotives clouding the air. Every few minutes the blast of a conductor's whistle and the slamming of carriage doors rattled the sleepy Virginians as they waited for their train to be called. When they heard the announcement for their train, Rolston sent Jane ahead to secure seats, while he made sure the trunks were loaded. He later caught up with Jane proudly sitting in a fine car with ample space and elegant fittings. Perhaps more aware of their poverty, Rolston pulled Jane out of first class and back to the third-class seats where they were ticketed. Jane remembers the famed Flying Scotsman train pulling out of King's Cross Station on the instant the station clock struck ten.

Rolston's supervisor during his two years at Edinburgh's New College was the Scottish theologian Thomas F. Torrance. Torrance had a reputation at the time for his theological scholarship on the church fathers and his work in the ecumenical movement. Almost as soon as he met Torrance, Rolston knew the two would have their differences. As a follower of Karl Barth, Torrance embraced the prevailing Protestant position that scripture, not nature, was the sole source of insight into God. He actively discouraged his students from the idea that the study of the natural world had anything to say about the character of the divine. Though Rolston was impressed by Torrance's rigorous approach to scholarship and learned a lesson from the strength of his scholarly convictions, he was never much drawn to his advisor's theological positions. Torrance gave every sign of being ambivalent about those of his American advisee.

Years after Rolston had left Edinburgh, his former supervisor became known for his work on the relation between theology and the natural sciences. Torrance claimed, controversially, that theology was really a kind of "theological science," making comparisons between methods in theology and in physics, and later winning the Templeton Prize for this work. Rolston was exposed to Torrance too early to benefit from the future interests of his advisor. In any case, Torrance's observations were mostly methodological. He never had much to say about the natural world itself, an omission Rolston would probably not have been able to forgive.

Rolston's research in Edinburgh ended up being focused on an area of traditional theological scholarship. His dissertation, later published by John Knox Press as *John Calvin Versus the Westminster Confession*, proposed the rejection of predestination and the recovery of grace in the Reformed Church. He challenged the doctrine of predestination, especially the doctrine of "double predestination," the view that those who were saved and those who were damned were chosen by God before their birth. The idea that everything was so scripted ahead of time was simply not plausible to Rolston. In his eyes, there was too much storied eventfulness in life. Contingency and chance were inescapable features of the order of things. Though lawlike in some respects, life contained both fixity and fluidity. Double predestination seemed to Rolston to squeeze the important elements of chance from history.

Though he rejected double predestination, one of the key Protestant commitments Rolston embraced in Edinburgh was the

Calvinist emphasis on grace. "My ancestors were wrong about predestination," he noted. "But the original reformers did have one thing right: Life is a kind of gift." As he watched the pipe bands play at the Edinburgh festivals, he noted how the Scots were equally attached to two anthems: "Scotland the Brave" and "Amazing Grace." The centuries-old struggle for independence against the more powerful English and the harsh realities of life in the rugged northern landscape had instilled bravery into the Scots. Rolston felt that more than any other people, the Scots seemed to fully understand how life involved struggle.

At the same time as acknowledging the importance of bravery in the face of unrelenting struggle, the early Scottish reformers had also insisted that God's grace played a crucial role in human affairs. Rolston's time in Edinburgh convinced him the Calvinists were right on this one. Grace was a recurring feature of life. Without God's providence, it would be almost impossible to make it through the struggles and the pain one encountered. The hardworking and devout Scots seemed to have figured out how to acknowledge the role of both struggle and divine providence in their lives.

Using what his experience told him to challenge what the theologians in his denomination professed, Rolston was confronting the age-old tension between faith and reason. On the topic of double predestination, faith and reason were starkly at odds. Faith embraced it; reasoned observation about the way the world worked made it seem unlikely. As Rolston wandered down Princes Street in the depths of the Edinburgh winter, he heard the ghosts of two of the city's most famous sons, enlightenment philosopher David Hume and Calvinist theologian John Knox, articulating the merits of each position to their supporters. Rolston spent a good deal of his spare time puzzling over how to balance the two types of explanation. One of his favorite spots to brood became Hume's windy grave high on Calton Hill, indicating perhaps that Rolston was already leaning more toward the empiricist side of the conflict than he was to the Calvinist. Even though reason caused him to reject predestination in his PhD thesis, faith drove him to hang onto grace, a commitment that not only lingered in his theology but eventually migrated to other parts of his thinking.

Though studying hard for his doctorate, Rolston was unable to set aside his curiosity for the natural world. The wilder parts of Scotland frequently drew him from town and he explored the highlands and glens at every opportunity. He generally hiked alone, spending many of his weekends rambling solo through heather and pine forests, climbing Ben Nevis and several other Scottish Munros. He became proficient at identifying the flora and fauna of the region and took lessons from Torrance about fly casting on the Scottish lochs. He got used to cold rain dripping down his neck and the importance of a steaming cup of tea in front of a coal fire on his return.

Rolston also made time to travel in Europe. In 1958, with the cold war at its height, he journeyed to Moscow during a break from classes. There he watched thousands of jubilant Muscovites march in a huge parade to celebrate the 1917 revolution. A balloon belching red smoke and depicting the Sputnik rockets sailed over the crowd. Given all the anti-Soviet propaganda he had been exposed to in the United States, he was surprised to see genuinely happy Russians celebrating their nation and its technological prowess. Rolston and his small group from Edinburgh joined the throngs of people waiting in line in Red Square to get a view of the embalmed corpses of Lenin and Stalin. The openness and humor of the Russians he met on the streets surprised him. They joked about waiting in long lines to see Lenin and Stalin because they wanted to make sure the two were dead. He went to a service at a Baptist church and was moved when the congregation gave a personal welcome to their American visitors. As in his earlier encounter with German prisoners-of-war in Charlotte, seeing the "enemy" face to face made him reflect on how strange it was that ideology could pull people away from their obvious biological commonality.

He and Jane used another intersession at Edinburgh to travel to Palestine not long after the Arab-Israeli war over the Suez Canal. There he was struck again by the role of suffering and struggle in shaping the history of a people. His group was compelled to visit

Jordan before visiting Israel. If they had tried it the other way around, the Jordanians would not have let them into the country with an Israeli stamp in their passports. Using his binoculars, their guide showed them the Israelis literally living in his own home, one he had been forced to flee. Crossing into Israel, the couple had to carry their own suitcases across a no-man's land one by one, with machine guns pointed at them from both directions. He remembers watching Jane's small frame making this walk in front of him, shaking slightly.

With his doctorate close to completion, Rolston and his wife prepared themselves to say good-bye to the town they had come to know by its nickname, "Auld Reekie." Jane took the boat back to New York while Rolston stayed for a few more weeks to put the finishing touches to his dissertation. He was feeling some nervousness about how his work was going to be received, given the slightly uncomfortable relationship with Torrance. His dissertation was a direct challenge to the key tenets of the Westminster Confession, the bedrock of Scottish Presbyterianism, and Torrance was a world-renowned professor of systematic theology. His worries proved unnecessary. After sending Jane a brief telegram—Thesis Accepted. Degree Awarded—Rolston flew back to the United States on one of the recently inaugurated jet services across the Atlantic and joined his wife.

Back in the southeastern United States, PhD in hand, Rolston saw everything swiftly fall into place according to the rough plan he had in mind. Returning once again to their family roots, Rolston and Jane stayed with their parents for a few months while he taught for one semester at Hampden-Sydney College, his grandfather's alma mater, forty miles from Richmond. With the semester completed, he was offered a position as pastor of High Point Presbyterian Church near Bristol at the southern end of the Valley of Virginia. Moving into a comfortable rural manse in the spring of 1959, he and Jane quickly settled into the routine of life at a country church, just as two other generations of his family had done. When he preached his first Sunday sermon at High Point, Rolston could not help feel a sense of enormous satisfaction and pride that he had returned to the landscapes of his childhood embarking on the career chosen by his father and grandfather before him.

He still had in the back of his mind the idea that he would move on to teach theology at a church college or seminary in the South at some point, but there was no urgency to pursue this path just yet. The story line he had managed to script for himself seemed to be working out just fine. Apparently forgotten for now, at least by all surface appearances, were the young man's scholarly interests in the natural sciences. To those on the outside, it was obvious that this well-educated, up-and-coming young theologian, recently finished with his doctoral research in Edinburgh, had better things to do than to go around gathering plants and insects. Why on earth would a pastor and theologian of Rolston's caliber and training want to spend his time looking down microscopes with the likes of Buggy Daggy?

Rolston's career was set. Or so it seemed.

PART II

A Pastor Gone Wild

4

Preacher and Appalachian Naturalist

> Christians and Jews do not turn to the Bible to learn natural science. Four centuries of developing modern science underscore what Galileo insisted with the launching of astronomy: The Bible teaches how to go to heaven, not how the heavens go. [*1996*]

AT FIRST GLANCE, THE ROLSTONS were a near-perfect fit for their role at High Point Presbyterian Church. The family history meant people in the valley knew the Rolston name. Between them, the couple possessed a rich knowledge of Virginia life. His degree from Davidson and his training at Union Theological and at Edinburgh left no doubt about the academic preparation he brought to the job. His calm demeanor enabled him to get along with just about anyone in the parish. He approached his clerical tasks diligently and was a capable spiritual guide. A casual bystander could hardly imagine the role of rural Virginia pastor sitting more comfortably on any person's shoulders.

Hidden within the polished exterior, a number of factors began almost immediately to eat away at the apparent harmony. Rolston retained from his younger years a deep and open curiosity about the natural world. Being back in Virginia resurrected powerful memories of time spent rambling through the woods as a child. As much as he aspired to be a great minister, Rolston found himself at least as interested in the complexities of his local ecology as he was in complexities of his human flock. This curiosity consistently drew him away from the church and into the enveloping landscape.

Most of his congregation at High Point probably did not care or even notice their pastor's interest in the natural world. Some doubtless looked positively upon his interests. A large number of the congregation were poor farmers. The idea Rolston knew a thing or two about the birds and the plants that shared their valley was not a problem. In fact, it was almost a valley tradition for reverends to know such things. It lubricated the social interaction on the steps of the church after Sunday services. Parishioners got into the habit of telling Rolston about the birds they noticed arriving as the seasons changed or the buds emerging on the trees. Some invited him down to their farms to identify unknown plants or insects, others took him coon hunting.

The problem with Rolston's interest in the landscape was not the knowledge of the natural world itself. It was the modern, scientific attitude with which he increasingly approached the world around him. Having had his eyes opened to the study of nature at Davidson, Rolston's approach to the flora and fauna of the Valley of Virginia differed radically from that of his parishioners. The pastor's way of talking increasingly made it sound like the scientific descriptions of nature had as much authority as the biblical texts. Where his pastorate saw only the blessing of God's bounty in the fields and woods, Rolston saw symbiotic relationships, ecological dependencies, and evolutionary lines of species and subspecies. His parishioners looked out from the door of their church on a fallen creation that needed redeeming through their hard work and prayer. Rolston looked out upon a complex ecology crying out for rigorous scientific classification and laboratory study.

The Christian God was by no means entirely absent from the accounts of the natural world Rolston formulated. The pastor was still committed to the idea of a creator God and to divinely bestowed responsibilities for humanity within creation. But as time went on, it became increasingly difficult for his parishioners to know exactly where their Creator, as they understood him, fit in. The first chapter of Genesis, with its six days of creation, was hard for them to reconcile with the things their pastor talked about. Terms like "subspecies," "ecological niche," and the new term the pastor was using—"DNA"—had not appeared in any Bible they knew. Some of them even heard him use the word "evolution," a term that to their mind had no place in a southern church. Rolston and his flock increasingly gazed out upon two different worlds, described in two different languages. The differences began to irritate his parishioners. From the pulpit, Rolston noticed more and more disgruntled looks as he preached. Over the weeks and months, he saw spaces beginning to open up in the pews. On the church steps after Sunday service, fewer of the farmers stuck around to talk.

In addition to the issue of the scientific language Rolston used, there was a second factor starting to alienate him from his parishioners. Rolston's extensive explorations of the Virginia countryside had made him increasingly worried about the destruction being caused in the valley by mechanized agriculture and development. It was in this valley, close to Rolston's birthplace, that Cyrus McCormick had developed the world's first mechanical reaper in 1834. The cascade of changes in farming techniques precipitated by McCormick's invention had led to massive transformations in the rural landscape. Hedgerows were being ripped out, streams redirected, and "pests" like foxes and raccoons were being trapped and poisoned. Rolston observed with increasing alarm the high cost of these common local practices. When he started to slip mention of some of these worries into casual conversations after church and on the local farms, his congregation balked. Not only did the pastor appear to look at the world through scientific rather than godly eyes, he also seemed to frown upon what they did in their working lives.

The congregation believed God had created humans in his own image, instructing them to "Have dominion over the earth," "Till the garden and keep it," and "Make the land flow with milk and honey." Clearing the land and making it bountiful was part of God's will. Farming was a way to express their devotion by making his garden productive. Uncultivated land was simply a waste. This assumption was not unique to the valley, it had guided the majority of Protestant immigrants since their arrival in the New World. It hardly seemed possible to the farmers that they might be mistaken. Both biblical and economic sense were on their side. The last person they needed to hear questioning the way they made their living was the pastor. The more opinionated among them started to talk openly about their dissatisfaction with Rolston.

As the months went by, Rolston felt more and more at odds with his congregation. Their steadfast refusal to view the world scientifically or to embrace new ways of thinking irritated him. He found himself unable to tolerate their enthusiasm for superstition and folk medicine. Their inability to see any biblical obligation to protect the woods astonished him. The underlying tensions were exacerbated when Rolston began to spend increasing amounts of time at another church where he ran the occasional service, a more liberal and less rural parish a few miles away at Walnut Grove.

As the rumblings of discontent at High Point escalated, several parishioners approached the church elders to ask for Rolston's removal. A number of closed-door discussions ensued. Not all the elders wanted to see him go but the majority decided they had no option. Enough damage had been done that they decided to honor the parishioners' request. Two of the elders more sympathetic to Rolston came around to his house and told him they would be going to the presbytery to pass on the congregation's wishes. The pastor was simply too out of touch with the people he was supposed to be serving. With the resentment now out in the open, Rolston quickly stepped down from his position. For probably the first time in his life, the Shenandoah Valley native had to come to terms with a sense of failure.

Fortunately for Rolston, there was little time to wallow in self-pity. He quickly took up a full-time post at the Walnut Grove church where he had occasionally been running the service. The new parish immediately suited him better. The congregation was a little more educated and more willing to accept the scientific talk Rolston brought into the pulpit and into his everyday conversations. Many of them welcomed a bright, young pastor who seemed like he could help lead them into modern times.

Ironically, given his ousting from High Point, Rolston was not progressive on all matters. On some social issues he remained deeply conservative. His opposition to interracial marriage and his position against the ordination of women remained firm for many years (though he later confessed to being more ashamed of voting against the ordination of women than of any other vote he cast in his life). He showed no flexibility in his belief that homosexuality was both sinful and intolerable. He found many of the more progressive currents in the church annoying. His father, still with the Board of Christian Education in Richmond and more liberal in his thinking than his son, spent hours trying to persuade him of the inevitability of change and the need to soften his positions. Despite this being the 1960s, Rolston was too much of a social conservative to give up his positions.

And yet when the issue was natural science, the pastor was on the cutting edge. He not only knew the intricate workings of the biological world but was also able to converse with authority about the mysteries of quantum particles and the relativity of space–time. A few of the congregation were still uncomfortable hearing their pastor hold forth in the language of modern science. They left Walnut Grove to find a pastor who talked a more familiar talk. Most of the congregation stayed and some even brought their friends to listen to the ways this unusual pastor tied the promises of the future to the familiar biblical stories. The church at Walnut Grove grew considerably under Rolston's guidance. One of his lasting achievements there was to manage the construction of a new church building to accommodate his growing pastorate's needs.

Despite the relative success at the new church, his popularity with the parishioners was an increasingly peripheral concern in Rolston's mind. More dramatic personal transformations were occurring within. The firing from High Point appeared to have nudged him onto a different path. By the mid-1960s he was spending one full day and several evenings each week prowling the fields and woods with his hand lens, identifying the birds, checking for tracks, and becoming familiar with the habits of the creatures who shared his home. Local farmwives began to notice a lean man with thick-rimmed spectacles spending long periods of time in the bushes staring at plants and insects. On at least one occasion, the police were about to be summoned to deal with a vagrant lingering near people's backyards when someone informed the worried parent it was just the Reverend Rolston looking at plants.

Increasingly frequently the pastor was nowhere to be found when his parishioners tried to reach him. Some times he was away backpacking on the nearby Appalachian Trail; at others, he had driven the fifty miles across the state line to Johnson City for classes at East Tennessee State University. Rolston's hunger for the natural sciences had rapidly turned into an obsession. Natural theology may still have been frowned upon by the Presbyterian establishment, but nature certainly seemed more interesting to Rolston at this point in his life than endless debates about predestination. At East Tennessee State, he sought out classes in anything that would make him better informed about the natural world. He took courses in evolutionary biology, botany, zoology, entomology, mineralogy, paleontology, meteorology, and climatology. The biology instructor, Jonas Barclay, himself a preacher's son, worked hard to provide Rolston with the knowledge his pupil craved. The geology professor, Lyman O. Williams, was delighted to have such a serious student in his class. The faculty became attached to the enthusiastic pastor from across the state line.

Before long, Rolston knew as much about the native plants as his professors. He was genuinely good at this stuff, driven by the sense deep down that the biological sciences had something profound to tell him about his place on earth. Increasingly, Rolston's hobby was becoming his vocation. He chased down opportunities to abscond from his job as a pastor and moonlight as a scientist.

One of his interests was birds. He befriended a local ornithologist named T. W. Finucane and became a regular observer of the fall hawk migrations along the Appalachian chain. The topography of the Virginia mountains funneled hawks over particular ridgetops on their way to Central and South America for the winter. One particularly tight funnel passed by the Mendota fire tower atop Clinch Mountain just forty miles north of Walnut Grove. In the fall, Rolston could be found lying flat on his back with binoculars pressed to his eyes counting the stocky silhouettes of broad-winged hawks migrating overhead. A few feet away, Mr. Finucane was on his back doing his own count. At the end of each flight of hawks they cross-checked their numbers and added them to the day's running total. In the course of an afternoon, two thousand or more hawks might sail over the old fire tower, the broadwings joined by a few wandering redtails, ospreys, sharpshins, and Cooper's hawks. Occasionally a golden eagle or a peregrine falcon would join the winged parade.

During these long days of sky gazing, Rolston mostly assumed the mantle of careful scientific observer, his currency no longer biblical passages but species and number counts. When there was a lag in the flights, the hawk watching offered ample opportunity for him to reflect on God's handiwork. Lying on his back gazing toward the heavens, he found the miracle of the migratory worlds of hawks directing his mind toward the Creator. He would rehearse passages of scripture under his breath: "Praise the Lord from the earth. . . . Mountains and all hills, fruit trees and all cedars! Wild animals and all cattle, creeping things and flying birds" (Psalm 148:7–10).

It was not only the hawks. From the same fire tower, Rolston observed clusters of monarch butterflies heading in the same direction as the broadwings on their own pilgrimage to southern breeding grounds. On Clinch Mountain he felt as if he were a personal witness to the fecundity and delicate complexity the Lord had set forth on earth.

Self-taught with the use of field guides, Rolston had become meticulous about recording his thoughts and observations on paper. He kept notes of all the trips he took, collections he referred to as his "trail log." Following in his father's footsteps, he sometimes turned these observations into larger reflective essays. He had already published accounts of his time in Scotland and in Russia in church newsletters and local papers. An account of a trip down the Grand Canyon had recently made it into the *Bristol Herald Courier* under the banner "Bristolian Shoots Rapids on America's Wildest River." People seemed to enjoy reading about the wanderings of the local pastor with a taste for outdoor adventure.

As his confidence in his writing mounted, Rolston added a few paragraphs of philosophical and religious musings to each of his essays. In the concluding paragraph of "September Hawking on Clinch Mountain," an essay he published in *Virginia Wildlife* in 1964, Rolston reflected on the flight of hawks.

> Each hawk silently, solemnly searches out his path in the trackless Appalachian sky. Surely there are things whereof they know but I am unaware. It is a most unique and, to me, a most satisfying recreation to escape the pressures and stress of my vocation by retiring to a solitary mountain tower to note these things so little noticed by the common rush of men. There is a mystery and majesty about the wild hawk in the wind-swept sky that elevates and frees the human spirit as well. "I do not understand: the way of an eagle in the sky." (Proverbs 30:19)

Theology and biology had always been connected in Rolston's mind. All he had to do was to convince his church to see things the same way.

Though somewhat rare in modern times, the phenomenon of a pastor with a fascination for natural science has extensive precedent in church history. In the sixteen and seventeen hundreds in Europe, few possessed an education as advanced as the local preacher. Clergymen trained in renowned centers of learning such as Rome, Oxford, and Paris. As part of their job, they traveled back and forth to the cities and towns and stayed informed about the larger political and intellectual currents of their time, as well as the scientific ones. Preachers maintained their authority over their parishioners in part through their superior intellectual preparation. Presbyterians were especially proud of their well-educated pastors.

Historically, the clergy differed from their parishioners in having a lifestyle that enabled continued intellectual development and reflection. Not bound to plowing the fields and bringing in the harvest and sustained by tithes paid by parishioners or wealthy benefactors, the clergy had the luxury of reflective time to work on their message and think in depth about theological issues. Hours spent walking through the countryside planning the Sunday sermon were an accepted part of the job. The natural world was deemed a suitable place for theological reflection. After giving up medicine as his first choice of career, Charles Darwin—despite nagging doubts about his theological beliefs—studied briefly for a career in the church, regarding it as a promising way to indulge a fascination for sea mats and sponges, a passion he had picked up in Edinburgh from a mentor named Robert Grant.

The classic mold for the pastor-naturalist had been set by England's Gilbert White (1720–1793). White was the curate of an Anglican church in the village of Selborne in Hampshire. He spent many hours ambling down the village lanes or moving slowly through his garden at a parsonage known as the Wakes. White was a keen observer and note taker. His meticulous details of bird and insect behaviors relayed in letters to his friends Thomas Pennant and Daines Barrington, and later published in his book *The Natural History and Antiquities of Selborne* (1788), revealed White's careful eye for the natural world. As was the case two centuries later with Rolston, White found nature to be not just a source of Christian wonder but also an object of serious scientific study.

White recorded in impressive detail even the most fleetingly glimpsed creature. He observed, for example, of the diminutive harvest mouse:

> From the colour, shape, size, and manner of nesting, I make no doubt but that the species is nondescript [that is, not known to science]. They are much smaller and more slender . . . and have more of the squirrel or dormouse colour. . . . They never enter into houses; are carried into ricks and barns with the sheaves; abound in harvest, and build their nests amidst the straws of the corn above the ground, and sometimes in thistles.

Living a hundred years before Ernst Haeckel coined the word "ecology," the curate recognized that there existed an organized interdependence to nature. Each creature possessed a distinctive and important fit in its environment. About earthworms, he noted, "though in appearance a small and despicable link in the chain of nature, yet, if lost, would make a lamentable chasm . . . worms seem to be the great promoters of vegetation, which would proceed but lamely without them." Through careful observation, White discovered scripted patterns at which he could only marvel. The explanation for these patterns, according to White, lay unproblematically in the hands of God.

Two hundred years after White's death and a continent away, Rolston sat squarely at the center of the tradition made famous by the curate of Selborne. There were American role models too. One was Elisha Mitchell (1793–1857), a theologian and geologist at the University of North Carolina, celebrated for his explorations of the Appalachians. Mitchell was the first to calculate that a peak (named after him), Mount Mitchell, was the highest point east of the Rockies. Mitchell later fell to his death exploring these then little-known mountains. Rolston was on the summit of Mount Mitchell dozens of times, never without thinking of its eponym.

Compared to his predecessors, Rolston had a much more sophisticated set of scientific explanations at his disposal to describe what he saw. In addition to his studies at East Tennessee State, he had made numerous trips to the University of Tennessee at Knoxville to meet with ecologists, to use the research library, and to update his knowledge. He had thoroughly annotated his copy of *Gray's Manual of Botany* and had read the correspondence between Darwin and Christian biologists such as Asa Gray. Biological staples such as Darwinian theory and Mendelian genetics, and emerging concepts such as trophic pyramid and symbiosis, enabled Rolston to explain phenomena about which White had to remain silent. Yet even with the increasing sophistication of the explanations, the impressiveness of creation to a theologically minded naturalist remained the same.

Though the pastor-naturalist has numerous well-known precedents, Rolston added the additional twist of being also a pastor-environmentalist. "One of the penalties of an ecological education," Aldo Leopold wrote, "is that one lives alone in a world of wounds." While White and Mitchell had lived in a mostly preindustrial age, Rolston lived in post–World War II America. When witnessing the destruction taking place around him in the natural world, Rolston felt a growing call to advocate for its protection. The problem he faced was that the religious tradition he occupied was decidedly ambivalent about this environmental message.

With two graduate degrees in theology, Rolston was well aware that the connection between the creator God and the call to protect the earth had a decidedly mixed history in Christianity. The Hebrew texts initially seemed to suggest that nature played a significant role in God's promise to humans. The covenant of creation in the book of Genesis bestowed upon humanity a blessed land. On the fifth day of creation, God declared: "Let the earth put forth vegetation, plants yielding seed, and fruit trees bearing fruit in which is their seed, each according to its kind" (Genesis 1:11). The garden of Eden was pronounced "good" in God's eyes. He gave to humans the plants and the animals for food and clothing but also provided the green plants as sustenance to "every beast of the earth, and to every bird of the air, and to everything that creeps on the earth, everything that has the breath of life" (Genesis 1:30). While only humans were made in the image of God, certain rules appeared to have been established for how people should treat creation. There was a divinely bestowed natural order to protect. Man was put in the garden of Eden to "till and keep it" (Genesis 2:15).

Despite the appearance of a blessed and productive earth in the original covenant, Rolston knew that many Christians read Genesis as setting up a distinctly hostile relationship between humans and the earth. The creation story appeared to strike several cruel blows against the idea that nature had moral significance for Christians.

At the conclusion of the creation narrative, some saw God setting humanity above the rest of nature in the role of lord and master:

> Let us make man in our image, after our likeness: and let them have dominion over the fish of the sea, and over the fowl of the air, and over the cattle, and over all the earth and over every creeping thing that creeps upon the earth. . . . Be fruitful, multiply, fill the earth and conquer it. Be masters of the fish of the sea, the birds of heaven and all living animals on the earth. (Genesis 1:26, 28)

The command to "conquer" the earth did not sound like a message about environmental protection.

Any remaining hope that humans might enjoy a benign relationship with the earth appeared to vanish when Adam and Eve were cast out of the garden of Eden. As punishment for their eating the forbidden fruit, earth had its blessed status stripped away. In his anger at Adam's disobedience, God decreed, "Cursed is the ground because of you; In toil you shall eat of it all the days of your life. Both thorns and thistles it shall grow for you" (Genesis 3:17–18). The descendants of Adam and Eve would in future gain sustenance only through hard work and struggle. The land no longer inspired gratitude for its fecundity, but resentment for its hostility to human life.

This struggle faced by humans appeared to come with an increasingly unqualified sense of entitlement to the fruits of their labors. When Noah emerged from the ark after the flood, God assured him that humans occupied a position of awful superiority over all other living beings.

> The fear of you and the dread of you shall be upon every beast of the earth, and upon every bird of the air, upon everything that creeps on the ground and all the fish of the sea; into your hand they are delivered. Every moving thing that lives shall be food for you; and as I gave you the green plants, I give you everything. (Genesis 9:2–3)

While Rolston still thought the Old Testament sent a message about good stewardship of the earth, he could see how certain passages lent themselves to conflicting interpretations about the proper relationship of humans to the environment.

These mixed messages had prevailed for at least the first millennium of Christian history. The desert fathers entered the wilderness expressly because of its barrenness, hostility, and the challenges it presented. Saint Francis of Assisi, on the other hand, embraced all of nature, including wild animals, for its richness and fecundity.

The continuing ambiguity about how Christians should regard the natural environment was well illustrated in the diaries written by fourteenth-century poet and Christian humanist Francesco Petrarch (1304–1374). Petrarch found that the spectacular scenery of the Italian and French Alps both impressed him and confused him about his loyalties. He had long wanted to scale Mount Ventoux, the peak dominating the horizon in front of his home. In one of Europe's first accounts of recreational hiking Petrarch told of recruiting his brother to come with him on the grueling all-day trip. As they climbed toward the upper reaches of the mountain, Petrarch found himself progressively more entranced at the scenes of splendor opening up before them. He recounted the experience to his friend Dionisio da Borgo:

> At first, owing to the unaccustomed quality of the air and the effect of the great sweep of view spread out before me, I stood like one dazed. I beheld the clouds under our feet, and what I had read of Athos and Olympus seemed less incredible as I myself witnessed the same things from a mountain of less fame.

Petrarch's exuberance at the wonders before him was brief. Close to the top of Ventoux, a sharp change of mood occurred. He recalled in detail how the change in sentiment came about.

> While I was thus dividing my thoughts, now turning my attention to some terrestrial object that lay before me, now raising my soul, as I had done my body, to higher planes, it occurred to me to look into my copy of St. Augustine's *Confessions*. . . . Now it chanced that the tenth book presented itself. My brother, waiting to hear something of St. Augustine's from my lips, stood attentively by. I call him, and God too, to witness that where I first fixed my eyes it was written: "And men go about to wonder at the heights of the mountains, and the mighty waves of the sea, and the wide sweep of rivers, and the

> circuit of the ocean, and the revolution of the stars, but themselves they consider not."

The text jolted Petrarch out of his reveries.

> I thought in silence of the lack of good counsel in us mortals, who neglect what is noblest in ourselves, scatter our energies in all directions, and waste ourselves in a vain show, because we look about us for what is to be found only within. . . . How many times, think you, did I turn back that day, to glance at the summit of the mountain which seemed scarcely a cubit high compared with the range of human contemplation—when it is not immersed in the foul mire of earth?

In an instant, the profanity of his enjoyment of the mountain became clear to Petrarch. How foolish to gaze at outward physical wonders when what mattered to God was the purity of the soul within. Petrarch concluded that hiking the mountain was a way to distract himself with trivial entertainments when he should be engaged in the more arduous task of preparing his soul for eternal life with God. The shame crushed him:

> I was abashed, and, asking my brother (who was anxious to hear more), not to annoy me, I closed the book, angry with myself that I should still be admiring earthly things who might long ago have learned from even the pagan philosophers that nothing is wonderful but the soul, which, when great itself, finds nothing great outside itself. Then, in truth, I was satisfied that I had seen enough of the mountain; I turned my inward eye upon myself, and from that time not a syllable fell from my lips until we reached the bottom again.

Petrarch had come to realize his true theological obligations. Good Christians should not allow themselves to get carried away with the majesties of the natural world, however beautiful they might appear. The soul was not of this world.

Back in the Valley of Virginia, scaling local mountains at every opportunity, watching eagles soar, and marveling at the surrounding beauty, Rolston pondered the real message of Christianity toward the natural environment. He knew most of his congregation would agree with Petrarch's turn inward toward the soul. Because of the obvious ambiguity in the biblical texts, he could not blame them for that.

He also knew it was becoming urgent that he persuade them they were wrong.

5

Leaving the Ministry

> Both science and religion are challenged by the environmental crisis; both to re-evaluate the natural world, and each to re-evaluate its dialogue with the other. Both are thrown into researching fundamental theory and practice in the face of an upheaval unprecedented in human history. . . . Life on Earth is in jeopardy owing to the behavior of one species, the only species that is either scientific or religious, the only species claiming the privilege as the "wise species," *Homo sapiens.* [*2006*]

IN CONTRAST TO MOST OF his congregation, Rolston had never thought the land was simply a resource for humans to use as they pleased. His Bible made a clear statement about God's ownership of the land; "the Earth is the Lord's and the fullness thereof" (Psalm 24:1). There was nothing ambiguous there. The scriptures also confirmed what he saw with his own eyes, that the Lord's creation was something of jaw-dropping richness and fertility. Deuteronomy did not exaggerate when it described a bountiful and sweet earth. Psalms were on the mark when they told of fertile pastures, mountains full of wildflowers, and rivers quenching nature's thirsts. Rather than emphasizing the Genesis passages that spoke of dominion, Rolston thought it more appropriate to pay attention to those declaring creation to be good. Neither Job, nor the writers of the wisdom literature, nor Jesus treated nature as if it were cursed.

Reading the Bible with a naturalist's eye, Rolston saw that God was committed to the continued health and fertility of the land he had created. Moses told the Israelites they would enter "a land of hills and valleys, which drinks water by the rain from heaven . . . a land which the Lord your God cares for" (Deuteronomy 11:11–12). Psalms similarly highlighted God's benevolent role.

> Thou dost cause the grass to grow for the cattle,
> and plants for man to cultivate,
> that he may bring forth food from the earth . . .
> The trees of the Lord are watered abundantly . . .
> the high mountains are for the wild goats;
> the rocks are a refuge for the badgers. . . .
> The young lions roar for their prey,
> seeking food from God. (Psalm 104:14–21)

Rolston recognized that in biblical times nobody knew much science by today's standards, but it appeared they could see what was going on around them. Jesus acknowledged the planet's fecundity when telling his disciples "the earth produces of itself" (Mark 4:28). Genesis even contained an evolutionary flavor, Rolston thought, when it talked about the seas "bringing forth" creatures in "swarms," a description the Virginia pastor noted was consistent with life's origin in the oceans.

Sensitized to read the Bible as both a Christian and a scientist, Rolston found abundant justification for taking care of the natural world. Earth was blessed with a complex ecology ordained and sustained by God. To doubt the theological significance of nature, as Petrarch had done, seemed to be a horrible misreading of the scriptures. Rolston too had climbed mountains with theological texts in hand. And when he reflected on phenomena like the fragile wing of the monarch butterfly that could yet carry it thousands of miles to breed, rather than doubt the value of creation, Rolston felt it more appropriate to hold up the Bible in two hands and exclaim with the psalmist, "O Lord, how manifold are thy works! In wisdom hast thou made them all; the earth is full of thy creatures" (Psalm 104:24).

But where Rolston found almost continuous theological inspiration on the summits of Clinch Mountain and in the forests around the base of Mount Mitchell, many in his religious tradition seemed to have long ago settled the question of nature's value in a notably different way.

Petrarch's equivocation on the summit of Mount Ventoux had captured what was at the time an unanswered question in Christianity. Should one look to the natural world to honor the creator or should one turn away from earth toward scripture and the soul within? The dominant theology of the medieval period initially seemed sympathetic to both sides of the issue. According to Thomas Aquinas, the natural world could indeed reveal important truths about the creator. Aquinas embraced the theological significance of the earth, noting Paul's claim in Romans that "his invisible nature, namely, his eternal power and deity, has been clearly perceived in the things that have been made" (Romans 1:20). For Aquinas, the earth was a legitimate source of revelation, the divine nature discernable through reflection on the natural world. Aquinas, in other words, was committed to the idea of a "natural theology," or what theologians sometimes called "general revelation." The twin pillars of later enlightenment thinking, science and reason, had a clear role in Aquinas's theology.

For human salvation to be assured, Aquinas knew it was also necessary to grasp truths available only through "special revelation." Special revelation involved turning one's attention away from the physical world and paying careful attention to the content of scripture. In contrast to general revelation, special revelation prized faith over reason, religious belief over empirical investigation. Recognizing the importance of both types of revelation, Aquinas worked hard to detail a system of theology in which faith and reason, special revelation and general revelation, could effectively complement each other.

The Protestant reformers, writing several centuries after Aquinas, switched the emphasis firmly back onto the scriptures and special revelation. Luther thought the Catholic Church had overstated how much reason could determine through contemplation of nature alone. The fall of man had corrupted both humanity and the natural world. It therefore seemed improbable that theological truths could be reached through corrupted man's reflection on a corrupted natural order. Knowledge of God lay *sola scriptura,* "in the scriptures alone." Redemption was *sola gratia,* "in grace alone." The natural world did not have much of a revelatory role to play.

After the Reformation, this trend away from natural theology continued, helped along by other rapidly moving cultural developments. The birth of modern science in the seventeenth century made it look increasingly unlikely that the workings of nature could reveal insights into divinity. The more physical and mechanical principles explained the natural world, the less need there was to invoke the will of a divine being. Science took the explanation of how nature worked out of the hands of theologians and put it into the hands of physicists and mathematicians. Nature had a mechanical and not a theological essence. These theoretical changes came with moral implications. A mechanical nature had less need for God's involvement. With God out of the picture, humanity was increasingly free to manipulate the physical world according to its own desires.

Though the divine role had lessened considerably by the seventeenth century, God did not disappear entirely from the mechanistic universe of the modern scientist. The image of God as a benevolent clockmaker building the machinery of the universe before winding it up and setting it in motion seemed to be an acceptable compromise between theological and scientific explanations. The divine clockmaker was ascribed varying degrees of ongoing responsibility for the operation of creation. The seventeenth century's most famous scientist, Sir Isaac Newton, insisted upon a fairly active and interventionist God, whose powers were continually necessary to keep the stars from collapsing in on themselves under the influence of gravity. Other scientists of the era, such as Pierre-Simon de Laplace, suggested that while God may have been necessary to create the universe, once this beautifully designed machine had been set in motion it simply ran itself according to physical laws.

With modern science progressively squeezing God from the day-to-day operations of the earth, there were increasingly few reasons to worry about human interference with these purely mechanical processes. The ambiguity Petrarch felt about how humans should regard the earth was rapidly being replaced by the idea that nature had little theological or moral significance. God cared about people, not passenger pigeons. Christians were told to "Love God" and "Love thy neighbor." They should work on saving their souls. Nobody was telling them to save the earth.

With the rise of capitalism in the nineteenth century, human domination of the natural order became an accepted Christian practice. Capitalism made the creation of wealth through the transformation of natural resources into its own end. This ideology lent itself particularly well to Reformation values. In *Protestantism and the Spirit of Capitalism,* Max Weber explained just how neatly these two world orientations fit together. The hard work and thrift of early capitalists offered a perfect display of Christian virtue. Individuals held the future in their own hands. Creative capitalists had divine sanction to convert nature into marketable products. The wealth generated could then be used charitably to reduce poverty and disease. If some of it was spent on luxuries and lavish lifestyles, nobody should begrudge the hardworking capitalists their success. The Bible often told of reward for those who followed God's laws. Industrially driven progress became part of humanity's manifest destiny.

The Genesis interpretation exhorting human dominion over the earth had by the end of the nineteenth century almost completely won out in Protestantism. One prominent historian of ideas claimed this made Christianity into "the most anthropocentric [human-centered] religion the world has ever seen." If earth was now theologically insignificant, the relationship between the natural sciences and theology could become largely one of indifference. There was no moral need for a Christian to know much about the natural world. Nature had little to reveal about God and there was certainly no point in wasting time trying to unite biology and theology. Some Christians still found nature beautiful and interesting. Some even found reason to praise God for creating it. It was generally agreed, however, that such thoughts were peripheral to Christianity's primary mission of saving souls. None of this bode well for the earth. And none of it bode well for Holmes Rolston III wrestling with these issues in a religiously conservative part of Virginia.

Rolston was convinced the "dominion interpretation" of Christianity's relationship to nature was mistaken. In addition to his own reading of the scriptures, he had deeply personal reasons to believe in the importance of God's earth. This was the third generation of his family to be raised in the rural Virginia landscape. He and Jane were thinking of starting a fourth. The stories told about his grandfather's love of this same valley half a century earlier and his memories of Will Long's home-style environmental ethic in Alabama suggested that his ancestors had always recognized the theological value of nature. Rolston reread the yellowed newspaper obituaries from his paternal grandfather's death in 1924. The *Staunton News Leader* had described the first Reverend Rolston as "particularly sensitive to the beauties of nature." The obituary added that, like Saint Francis, his grandfather had "felt his kinship to everything that breathed the breath of life." The *Christian Observer* told similarly of him "visiting God along the mountain streams." His grandson now watched with dismay as the same streams were filled with mine tailings or silt from logging operations. Slipping in and out of his roles as pastor, grandson, and environmentalist, Rolston knew that he had to persuade Christians to take an interest in ecology. But by this time the problem was more than simply ambivalence. In fact, for the last hundred years or so, biology and theology had hardly been on speaking terms.

The publication of Darwin's *Origin of Species* in 1859 transformed the relationship between Christian theology and the natural sciences from one of benign indifference to one of antagonism. Darwin's theory of the transmutation of species was utterly at odds with the biblical account of creation. Even though Darwin's theory fell short of explaining how life had originated on earth, it offered a plausible account of the life-forms now present. And the theory was clearly backed by the evidence. Not only was it consistent with what was known about animal morphology, embryology, and the geographic distribution of species, it also gave a compelling explanation of the fossils of extinct flora and fauna scientists were discovering almost daily in rock formations across the world. Darwin's theory of natural selection, a theory simultaneously proposed from Indonesia by another English naturalist named Alfred Wallace, called into question all previous accounts of life.

If the hypothesis of species descending through gradual modification from long-dead ancestors was not troubling enough for Christian orthodoxy, Darwin's theory had an even bigger surprise up its sleeve. Most threatening of all for Christians was the fact that Darwin's theory also provided a naturalistic account of the origin of humans. Humans, Darwin argued in the *Descent of Man*, were not created spontaneously in the image of God but evolved over time from primate ancestors. The hypothesis undercut the crucial Christian supposition that human origins were unique. The idea that people were not metaphysically special but had the same genesis as other animal species was a devastating threat to orthodoxy. Biology and theology were now virtually at war.

The battle lines were fortified almost immediately with a number of high-profile confrontations. Thomas Huxley's encounter with Samuel Wilberforce in Oxford in 1860 was one of the earliest and most famous. Huxley, who would later be known on both sides of the Atlantic as "Darwin's Bulldog" for his articulate defenses of evolutionary science, took on the bishop of Oxford at a meeting of the British Association for the Advancement of Science. According to various reports of the encounter, Huxley left the church's position on the spontaneous creation of species bloodied and bruised. This was not before Wilberforce had won laughter from the benches for portentously asking Huxley which side of his family—his grandfather's or his grandmother's—were descended from apes. Huxley's caustic reply—that he would rather be descended from apes than from a person of the cloth who refused to accept the lessons of science—illustrated the emerging bitterness. The drama in Oxford that day caused at least one delicate Victorian lady in the audience to faint.

Sixty-five years later on the other side of the Atlantic, Clarence Darrow and William Jennings Bryan squared off on the same issue in the "Monkey Trial" in Dayton, Tennessee. The State of Tennessee had decided to prosecute substitute teacher John Scopes under a recent statute banning the teaching of evolution in the schools. Darrow, a lawyer based in Chicago, came down to Dayton to defend Scopes's right to teach evolution. The trial rapidly turned into a media circus. Chimpanzees, allegedly to appear as witnesses for the prosecution, were paraded outside the courthouse to the amusement of onlookers. After eight days of the trial, Darrow lost the case on a technicality but won the publicity war. In what some people later claimed was a symbol of his pyrrhic victory, Bryan, who had been a popular legislator earlier in his life, died of natural causes five days after the verdict even before leaving Dayton. The effect of the Scopes trial was to add significantly to the polarization of the two sides. It left a particularly bad taste in the mouth of southern Christians who saw the trial as another attempt by northern intellectuals to impose unwelcome views on their former enemies.

The battle over human origins now became the central focus of the tension between Christianity and modern science. The earlier questions about the theological significance of nature and the possibility of natural theology had been subsumed by this more pressing debate. Taking a position on human origins was becoming almost definitive of one's credentials as a Christian. Suspicion of biology in general, and rejection of evolution in particular, was almost a requirement for church membership, particularly church membership in the South. It was Rolston's commitment to scientific biology—and, by reasonable inference, to evolutionary theory—that his congregation at High Point found so unnerving.

Forty years after the Scopes trial, Rolston could see that he had his work cut out for him. Before he could even begin to get Christians to care about the earth, he knew he would have to reduce the antagonism between evolutionary biology and Christian theology. Even Christians sympathetic to the idea that God wanted them to care for creation found it difficult to reconcile the language of the

Bible with Darwin's account of human evolution. From his own experience Rolston knew that evolution was far too inflammatory to discuss from a southern pulpit. His task was no longer as simple as following Gilbert White and using inspirational descriptions of nature to persuade Christians to respect the earth. Natural science and Christianity had first to be put back into conversation with each other. Rolston realized he would have to go several layers deeper than Gilbert White had gone. In fact, he would have to go several layers deeper than anyone had gone before.

Characteristically for a man who looked up to his ancestors, Rolston began his search for reconciliation by looking carefully at the theological resources closest to home. He wondered whether there was any room in Presbyterianism to buck the antievolution trend. He was educated enough as a scientist to know that the major theories were constantly changing. Modern Christians must therefore not tie themselves to a single scientific account but must look for a flexible marriage between science and religion. It struck him that the underlying Protestant commitment to continual reform, *semper reformanda,* should prepare them well for such flexibility. This seemed like a good start, at least in the abstract. Was there anything else he could find in Protestantism that worked in his favor?

The emphasis in the Reformed Church on the importance of a personal relationship with God initially suggested a yawning doctrinal chasm between Protestants and the natural world. The earth was not particularly involved in whatever relationship individuals had with their God. The distance Protestants deliberately sought to create from Catholic natural theology only added to the estrangement. In the mid-twentieth century, Calvinist giant Karl Barth had explicitly and emphatically confirmed the rejection of natural theology when he declared *Nein!* to theological reflection on nature in one of his book titles. Rolston's PhD thesis supervisor in the 1950s at Edinburgh, Tom Torrance, had fully bought into Barth's rejection of nature.

On careful scrutiny, however, Rolston found several elements of Calvinism that, despite his denomination's poor reputation among environmentalists, indicated the potential for a softer relationship to the natural world. Take, for example, the notion of grace. In his own doctoral dissertation, Rolston had highlighted the importance of grace in Calvinism to compensate for the diminished attention it received in the Westminster Confession. On the surface, the concept of grace did not look very helpful. Grace was the blessing that allows one's soul to be saved. It operated within the distinctive Protestant framework of the personal relationship between the individual and God. However, in addition to the "special grace" God bestowed upon particular sinners for their individual salvation, Calvin claimed that God also bestowed a "common grace" on the whole of creation. Referring to this grace as "spirit," Calvin declared in his *Institutes of Christian Religion*:

> The beauty of the universe (which we now perceive) owes its strength and preservation to the power of the Spirit . . . for it is Spirit who, everywhere diffused, sustains all things, causes them to grow and quickens them in heaven and in earth. . . . [I]n transfusing into all things his energy and breathing into them essence, life, and movement, he is indeed plainly divine.

The idea of God sustaining and quickening the natural world through common grace seemed, to Rolston, to be a move in the right direction.

As a result of the emphasis on common grace, Calvin shared with Catholics a willingness to look at the natural world for insight into God. He claimed there existed "innumerable evidences" of God's wisdom in creation:

> God not only sowed in men's minds that seed of religion of which we have spoken but revealed himself and daily discloses himself in the whole workmanship of the universe. As a consequence, men cannot open their eyes without being compelled to see him.

Calvin could turn rhapsodic about the beauties of creation, claiming that even the most ignorant see "more than enough of God's workmanship in creation to lead him to break forth in admiration of the Artificer." Rolston was mystified by how Barth could read this as *Nein!* to natural theology.

This openness to the study of nature led Calvin to refer to the universe as a "mirror," a "large book," and a "theatre" that displayed God's work. Rather than distrusting science's efforts to describe this work, Calvin explicitly championed it. He praised science as one of the "most excellent benefits of the divine spirit," one that ought to evoke our awe and gratitude. Calvin considered it not only a window into God's works but also an opportunity to find some humility in the face of earth's grandeur. The budding naturalist and environmentalist now preaching at Walnut Grove Presbyterian Church in Virginia could hardly have put the case better himself.

The way Rolston read Calvin, there were plenty of reasons to love both the gospel and the landscape. He knew that his own Scots Presbyterian ancestors had rejoiced in the natural wonders they discovered when they first settled in Pennsylvania, Virginia, and the Carolinas. Coming from the hauntingly beautiful but denuded mountains and glens of Scotland, they found themselves particularly moved and inspired by the fertile landscapes of their new home, seeing ample evidence of God's handiwork in creation. Rolston was confident that within his own tradition could be found some of what was needed to bring theology and ecology back into dialogue.

If one of the theological pieces came from his sympathetic rereading of Calvin's *Institutes*, other valuable insights came from his daily experiences in the woods. The time Rolston spent roaming through the forested mountains of Appalachia supplied further deep drafts of inspiration for tackling the puzzle in front of him. The mountains themselves seemed to point to something.

Mountain landscapes had always possessed significance in Christian scripture. Rolston remembered how one psalmist had noted, "as the mountains are round about Jerusalem, so the Lord is round about his people henceforth even for ever" (Psalm 125:2). Another psalmist had asked rhetorically: "I lift up my eyes to the hills. From whence does my help come" (Psalm 121:1). Rolston also noted in the book of Job: "he ranges the mountains as his pasture and he searches after every green thing" (Job 39:8). It was clear to Rolston that mountains and hills had always been biblically important. In an article he had recently written for the periodical *Christianity Today*, Rolston wondered out loud about the reasons for this confluence: "Do altitude and the beauty of the hills give highlanders a constant reminder of the Creator that those who dwell below have not? Perhaps it is easier with the Smokies on the horizon to sense the presence of him who once spoke at Mt. Sinai." He speculated later that there might be "something about the upward sweep [of the mountains] that brings the cosmic into focus."

Part of the significance of mountains in Christianity was the way they sustained two powerful metaphors. The first metaphor, which had its philosophical origins in Plato, suggested that knowledge of God could be achieved only after a laborious ascent from the lowlands of everyday life toward a higher plane. The divine had always been understood as located above the earth. "Elevating yourself" allowed for a better view of the heavens. Mountains provided physical and spiritual proximity to the divine provided one was willing to take on the task of lifting one's mind up and away from earthly things toward God. God's kingdom, the scriptures reminded us, was not of this world.

The metaphor of closeness to God from the mountaintop could be supplemented by a second equally potent metaphor of observing from on high the fecundity of creation below. From the mountain one could see the landscape laid out in all directions. Fields stretching out to the horizon bristling with crops, streams full of fish, and countless miles of sheltering forest filled with deer, bear, and foxes presented a rich spectacle of earth's plenitude to one's gaze. The elevation combined with the struggle of the climb to generate an appropriate vantage point from which to contemplate the fertility of creation beneath.

A Protestant theologian named Paul Santmire suggested that part of the reason Christians had neglected their obligations to the earth is that they had focused on the first image of ascent and looking upward beyond earth at the expense of the second image of looking downward toward earth's fecundity. Santmire recommended that emphasis on the contemplative downward gaze might better encourage Christians to appreciate the gift of God's earth.

The suggestion that Christian environmentalists should look down with thankfulness on creation as well as up toward the heavens would certainly have met with Rolston's approval. But having spent most of his life exploring Appalachia's mountain ecology, Rolston sensed that this was not the whole lesson Christians needed to learn from mountains. From his own experience of Appalachian ecosystems, Rolston knew that a full appreciation of mountain landscapes required more than simply a switch in metaphor in order to better appreciate the views below.

As his knowledge of natural science increased, Rolston had learned to approach mountain hikes a little differently. The highest point reached on the climb increasingly became only an incidental part of the experience. The journey itself was also central. He did not need a spectacular visual conclusion to enjoy a hike. Every couple of feet up the trail there were plants to identify, mosses to inspect, insects to examine, and animals tracks to study. The observational skills appropriate for the Appalachian ecology required as much crawling around on hands and knees as they did striding upward with one's eyes fixed on the summit. Rolston found himself using his hand lens as much as his binoculars. As the elevation changed, so did the ecosystem. The way he saw it, to walk straight past any of the intricate details was to miss many of the key aspects of creation. In the solitude of his Appalachian hikes, Rolston learned to appreciate the truth of the aphorism "You're not going somewhere; you're already there!"

It is not that Rolston had lost all interest in the views from the summit. On top of the mountains, he loved to inhale the clear air and search the horizon for signs of an early rising moon. He was always moved when he lowered his gaze and surveyed the fecundity of creation below and inspired when he lifted his eyes toward the heavens above. But as his ecological and theological sensibilities matured, Rolston increasingly found that on a mountain hike the summit was just one part of a long day of an enlivening and spiritual journey.

The importance of the journey pointed toward what he was beginning to see lay at the center of nature's value. Nature's real value was as something living and dynamic rather than something distant and magisterial. For Rolston, nature was theologically stimulating in the here and now. The traditional picture of God standing on high, creating and blessing the earth, and making a covenant with his people to be good steward was, at best, only part of the story. Christians needed to recognize the experience of rich and surprising encounters with nature available at every moment to the careful observer. There was grace to be experienced in the bloom of the wild iris and twitch of the deer's nostril, alert to danger, down by the creek at dusk. Thoughts of the divine were evoked as much by the ruffed grouse flushing from the blueberries on the trailside as by any reading of scripture near the mountaintop. On his hike up Mount Ventoux, Petrarch had simply looked in the wrong direction and missed much of what was going on. Rolston might have recommended Petrarch bring a field guide with him up Mount Ventoux rather than Augustine's *Confessions.*

Rolston recognized the nub of the problem for his Christian congregation. Nature's moral and religious value lay in its ongoing creativity. To foster any kind of appreciation for nature's spontaneity and complexity, Christians were first going to have to learn not to fear the science that could illuminate it. They needed to relearn how to have comfortable conversations with biologists in order to recognize the abundant theological values inherent in creation. Even as more and more of the intellectual pieces were beginning to click together for Rolston, the prospects for a rapprochement were looking increasingly bleak.

Scientific advances throughout the twentieth century had continued to turn the screw on traditional Christian teaching. The dawn of the space age in the 1950s suddenly made questions about the origin of the universe look much more answerable in the near future. The capacity to watch atomic particles at work with the electron microscope was for the first time providing real knowledge of the building blocks of matter. The discovery by Watson and Crick of the structure of the DNA molecule, "the secret of life" as they liked to call it, made possible explanations for biological phenomena that had previously lain entirely in the realm of the religious. The patenting of the birth control pill had given people increased control over the timing of human life. New thresholds of scientific explanation were being crossed every few months. On the other side of these thresholds were still more immense questions, a number of which cut to the very core of human self-understanding. As Rolston labored through the intellectual puzzles, he became acutely aware that the longer Christians remained uncomfortable with the sciences, the more their alienation would increase and the more the natural world was going to suffer at their hands. To view the physical and biological sciences and their associated technologies as hostile to Christian theology was an unnecessary and grave mistake. Rolston increasingly felt the urgency of doing something about it.

Throughout his years as a pastor, Rolston had always intended to step down some day to teach at a Christian college in the South. In those days, it was not uncommon for a talented pastor to be asked by one of the church colleges about any interest in joining a teaching faculty. The colleges all had numerous Bible and theology courses to be taught and they were constantly short of professors. With his father being known regionally for his work in Presbyterian theology and with Rolston himself having had such a distinguished academic past at Union Theological and at Edinburgh, he assumed he only needed to bide his time at Walnut Grove before an invitation would land in his lap. Much to his dismay, no such invitation had been forthcoming.

It was now the late 1960s and Rolston had already spent far longer as a pastor in the Valley of Virginia than he had anticipated. With the invitations to teach still failing to appear, he started to become demoralized. He began to apply actively to colleges he thought might want him, including Washington and Jefferson College in Pennsylvania, St Andrews in North Carolina, and his alma mater,

Davidson. He still did not get any offers. People he knew were getting those jobs. Rolston recalled the job he applied for in Laurinburg: "Didn't get it. Mac Doubles got it. Nice guy. Good friend of mine. Mac got that doggone job." Jane detected a rare hint of anxiety and frustration in her husband at home. Things were not quite slipping into place.

Physicist, theologian, pastor, and now biologist. Rolston was starting to feel alarm at how his interests were zigzagging across the map. Were the paths he traveled simply too scattered? He looked for comfort in a proverb often used by his Shenandoah Scots forebears: "God writes straight with crooked lines." He began to think about going back to school to get a different kind of graduate degree. Rolston knew he had to resolve the tension between science and theology, between genes and Genesis as he now saw it. To get at this problem, he felt he needed a different perspective. The path forward was not entirely visible from within theology, nor was it entirely visible from within the natural sciences. And it was certainly not visible from his current job as a pastor at a southern Presbyterian church.

To find the way forward, he needed somewhere else to stand. First he had to figure out exactly where that place was.

6

Discovering Philosophy

That we can be upset when lost depends upon a baseline emotion of being at home. Our homes are cultural places in their construction, but what we add again is that there is a natural foundation, a sense of belongingness to the landscape. [*1979*]

WITH THOUGHTS ABOUT THE FUTURE percolating uncomfortably through his mind, Rolston found both solace and inspiration in almost daily explorations of the Appalachian countryside. When he needed to settle his nerves, Rolston rarely sought out friends and company. He grabbed his boots, some guidebooks, and a couple of maps and took off to the woods.

Jane accompanied him on some of the hikes and over the years at Walnut Grove they took numerous camping trips together in the Smokies and on the Blue Ridge Parkway. Jane appreciated Virginia's abundant natural beauty and enjoyed even the arduous climbs with her husband. But she was also getting used to the idea that her husband was driven more by the need to study and understand the landscape than he was by the need for company. She knew in her heart she had married a loner. Most of the time, Rolston spent his days in the woods alone.

Often when he set out into the wilds, Rolston was startled by the degree of human impact on the landscape. He would find himself just a mile down the trail and discover that the backside of a mountain had been clear-cut, out of sight of the main roads. As he witnessed the transformations, the sense of urgency he felt to work for the protection of nature became stronger.

Now, in the late 1960s, the nation had started to wake up to the environmental cost of the previous decade of economic expansion. Widespread toxic waste and pollution had led to increasing national demands for protection of human health. The people were upset and politicians were starting to listen. Early versions of the Clean Air and Clean Water Acts had passed the United States Congress. Congress was talking about a national environmental policy act, legislation that would make environmental protection a key component of all government projects.

Rolston was heartened to see progress at the national level on environmental issues, but something about the way most of the political arguments were being made disturbed him. The laws were all directed at *human* health and *human* benefits. For Rolston, this focus on humans seemed to leave something out. He was convinced that there was moral and religious significance to nature independent of its contribution to human well-being. Nature was not there simply to make people feel good when they recreated. His grandfather, Holmes Rolston I, had found the creeks and hillsides of Virginia to be a sanctuary, not simply a sanatorium. Discovering that his grandfather's refuges were being polluted by mine tailings or hauled off to the nearest lumber mill offended both Rolston's sense of family history and his sense of nature's inherent goodness.

Rolston obviously needed no persuading about the theological grounds to care for the earth but through his firsthand encounters with Appalachian ecology he had progressively gained the conviction that a person did not need to rely on the scriptures to find a reason to protect nature. There were other compelling reasons to save nature, different even from the significant concerns about human health. It seemed to Rolston that the creative and prolific tendencies within nature itself—setting aside the question about whether God had created them—presented a moral reason to care. It was as if a patient naturalist could almost *see* the moral values when examining the pistil of the trillium or the gaping beaks of a woodpecker's downy young. Nature seemed to Rolston to be brimming over with remarkable biological achievements, achievements that science could play a vital role in illuminating. At this point in his life, he was fumbling his way toward an articulation of these thoughts about nature's value. It was already clear to him that a convincing secular argument for the value of nature would make a useful addition to the scriptural one already forming in his mind.

Rolston saw the need for a new path between ethics, science, and religion, but he knew he was not yet familiar enough with the territory to forge the path himself. He needed additional background. Since he already had two graduate degrees in theology along with ten years of pastoral work in Virginia, he was certainly not lacking the education in religious studies. Over the last few years, with the help of field guides and the classes he had taken at East Tennessee State University, he had also become a proficient self-taught naturalist. He did not need any more background in the natural sciences. As he started to contemplate the road ahead, a different avenue began to appeal to him. When humans first started asking themselves fundamental questions about the nature of the world around them two thousand years earlier in ancient Greece, they created a name for the overarching discipline. That name was "philosophy."

During his studies in Edinburgh over a decade earlier, Rolston had been impressed with the questions asked by philosophers. Tom Torrance and some of Rolston's other theological mentors had always emphasized faith over reason and had tried to downplay the importance of philosophy. Its methods, after all, did not require allegiance to any particular religious creed. Fortunately, one of his professors at Edinburgh, John Baillie, had worked hard to keep a sense for the value of philosophy alive in his students. The realizations Rolston had come to at seminary about the importance of interpretation and history were philosophical because they posed hard questions about the foundation upon which the truth claims of theology were made. The insights he had reached about the fluid meaning of words after learning Hebrew and Greek were likewise philosophical.

The power of philosophy's disciplined rational inquiry to investigate questions of truth and method was something Rolston knew he now needed for his quest. He secretly thanked Baillie for his efforts. Ten years later he could better appreciate how philosophy possessed a unique perspective from which to challenge the assumptions other disciplines took as their bedrock principles. To get science and religion into dialogue about nature he needed to come at them from a different angle, one that offered a critical perspective on both. Philosophy looked like it might provide that perspective.

Within the many subdisciplines of philosophy, Rolston found one particular area coming closest to the sort of issues he had in mind. With the country in the grip of the atomic age, the philosophy of science had recently become a hot area of research. This important branch of contemporary philosophy investigated the theoretical support structure on which the sciences operated. It asked questions about the ultimate makeup of the physical world and evaluated the framework scientists employed to investigate it. It probed reality in the depth Rolston would need. He decided to get an education in the philosophy of science.

Rolston's plan for his upcoming intellectual journey was now taking shape. He would leave his pastorate at Walnut Grove and get a graduate education in the philosophy of science. The one big hitch was that he had no coursework in any area of philosophy. To make matters worse, he had spent the last decade of his life as a rural pastor in Virginia. Philosophy departments were often suspicious of applicants with a religious background. A country preacher wanting to do philosophy of science would certainly raise eyebrows among graduate committees. The last thing they needed in philosophy of science was somebody quick to take a leap of faith or ready to accept truth on the basis of biblical authority.

On top of this general suspicion of religion, many philosophy departments throughout the Anglo-American academic world then embraced a school of thought called "logical positivism," in which only those things experimentally observed could be used as starting points for knowledge. The idea was to make no knowledge claims that did not originate in reliable observations. The paradigm case of a reliable observation was an observation made in a physics lab. For most logical positivists, speculations about nonphysical entities like divine beings or moral values were simply meaningless. They certainly could not be used as a basis for a belief system. None of this looked very hopeful for Rolston's plan to pursue a graduate education in philosophy.

Undeterred, the pastor quickly took an introduction to philosophy at a local college to get at least one course on his academic transcripts. He also started researching a suitable master's program in the philosophy of science. In the heyday of positivism, when academics talked about the philosophy of science, what they really meant was the philosophy of physics. Philosophy of biology hardly yet existed. Atomic particles seemed much more essential than genes or photosynthesis. Fortunately, Rolston knew enough about physics from his Davidson days to compose a decent application letter. He recognized his best bet would be to apply to a program specializing in the philosophy of physics and then tailor a course of study to suit his own interests.

During Advent of 1966, whenever he could carve out time from his busy clerical schedule, Rolston worked late into the night on his applications for graduate school. He sent dossiers to ten schools and was rejected by most of them. To his pleasant surprise, the University of Pittsburgh, which had one of the strongest programs he considered, accepted him. The following summer Rolston and his wife started packing their belongings for Pittsburgh.

While all these plans made sense in Rolston's head, what he had not anticipated was the effect they would have on his heart. The decision to go to graduate school meant a wrenching move from the beloved landscapes of both his and Jane's birth. He had been away from Virginia temporarily before, when his family moved to Charlotte during World War II and when he and Jane had lived in Edinburgh for two years during his doctoral studies. Nevertheless, Rolston had still spent the majority of his life in the Valley of Virginia. For the last decade, he had created an identity for himself as a Valley of Virginia pastor. Moving away from the landscape in which both his father and his grandfather had raised their families was a decision that kept him awake at night. He felt sick at the thought of leaving. He comforted himself with the idea that before long he would return to teach at a southern Christian college, but he realized that the decision to go to graduate school in philosophy was leading him into uncharted territory, both intellectually and geographically. Jane noticed a rare lack of confidence in her husband about the upcoming move. Sitting in a sandwich shop one day, trying to sort through a list of what furniture to take from their spacious manse to the small yellow brick house they had rented in Pittsburgh, Rolston threw up his hands in frustration. He confessed to Jane that she was adjusting much better to the idea of the impending transition than he was.

In part to give voice to his sadness, Rolston wrote about his departure in an emotive essay for *Virginia Wildlife* titled "Farewell, Washington County." The essay is a paean to the landscape he was leaving. Departures, he observed, evoke memories and memories often evoke a tribute. To pay that tribute, he looked back at some of the most memorable experiences over his years in the Valley of Virginia.

The essay recounted long and tranquil mornings spent studying the flora in fragrant woodlands and frightening moments on stormy mountaintops. He explained the wonder, later confirmed by a University of Tennessee bryologist, of discovering a species of moss never before seen in Virginia. Like Aldo Leopold in his *Sand County Almanac*, Rolston cataloged the distinctive passing of the seasons, drawing attention to the characteristic excitements of each month in the valley. He told of following tracks through the December snow to discover where the fox had wandered during the night. He recalled coming upon a frog-filled pond late one February at the top of an unexplored trail on the flanks of Holston Mountain. He recounted the memorable occasion he walked up on a covey of partridges on the side of the road, with their tails facing inward in a defensive circle and a dozen pairs of eyes that met his "in an encounter that somehow went right to the nerve of life itself." He closed the eulogy to his native landscape with a prayer acknowledging the mystery and majesty of the land he was leaving:

> What wonders and pleasures lie in the realm your boundaries enclose, from the majesty of Abrams Falls to the mystery of Ebbing Spring, from limestone sinks to green pastures and templed Hills. De Profundis! What abundance and splendor are compacted here! Lord, bid time and nature gently spare these hills that once were home.

When the day of their departure finally came, both Rolston and his wife drove away in tears. Jane later recalled her husband crying "all the way to Roanoke," at one point sobbing so hard he had to pull off the road to dry his eyes. It was with a great deal of trepidation that they found themselves later that day standing in front of a small brick house in Pittsburgh, not far from an elementary school where Jane had secured a job teaching in the third grade to support her husband's studies.

Rolston finished his degree at Pittsburgh within a year. As he suspected, his professors drilled him with a lot of hard-nosed philosophy of physics. At the same time, he also gained his first formal education in the methods of philosophy. He got an overview of the discipline's history, taking classes on Plato and Aristotle, and on the modern philosophies of Descartes, Spinoza, Hume, and Kant. He learned the contemporary analytic philosophy of W. V. O. Quine and Gilbert Ryle and got a feel for how Continental thinking was distinct from the Anglo-American philosophical tradition. Rolston studied the basics in ethical theory, learning for the first time in his life how to ground an ethic in something other than divine commands. This gave him more of a feel for how philosophers made their arguments and for how the relationship between philosophy and theology became progressively more strained during the Enlightenment. He learned in more detail why John Knox and David Hume had occupied such different portions of the cultural memory in Edinburgh.

In his studies of the philosophy of physics, Rolston recovered a taste of the excitement he had experienced at Davidson for how numbers captured the workings of the universe. At the same time, in the years since Davidson, his self-guided explorations of the natural world had confirmed his suspicion about the disappointing abstraction of physics and mathematics. Physics constructed progressively more elaborate theoretical bridges in relative detachment from the world of everyday experience. The dirt that worked its way under the fingernails of biologists and ecologists and got stuck in the soles of their boots was entirely absent from the physics lab. Ultimately it might be physics and chemistry that made those mushrooms picked in Mud Hollow near the High Point church glow in the dark, but neither physics nor chemistry explained why there just happened to be this particular fungus on the damp log beneath the hickory tree. The lack of down-to-earth explanations jarred against Rolston's instincts as a natural historian.

Abstraction from real life was not the only thing that bothered him about physics. Classical physics, he could now see, was highly deterministic. A world governed by physical law left little room for spontaneity and chance. Everything was fixed ahead of time. The idea that everything was law-governed reminded him of the doctrine of double predestination he had found so troubling in his Edinburgh days. It left no room for spontaneity. Rolston had become increasingly intrigued by how much was built upon chance in life. Why did it happen in postwar Charlotte that polio struck one young girl while leaving her brother untouched? Why did Rolston stumble on those luminous mushrooms in Mud Hollow, pushing on another ten minutes in his hike, when he had earlier started to turn around to get out of the woods by dark? For what reason did the fox miss the first rabbit he chased but catch the second, if the blood in the snow was to be believed? Natural history was full of a blessed storied eventfulness that physics completely bypassed. Chance did not simply *interrupt* history. It *made* history. Luck seemed to be as much a driving force in life as the inexorability of physical law.

Rolston knew from his studies at Davidson that quantum physics opened up more room for uncertainty and contingency than the classical physics of Newton had once allowed. This is partly why he found himself drawn to it. Despite the presence of indeterminacy in the quantum world, physics as a whole seemed inadequate to explain historical eventfulness. It did not leave room for the full story. Rolston became increasingly convinced that biology and ecology, with their recognition of the deeply historical character of life on earth, were much more explanatory than the philosophy of physics could ever be.

In formulating these thoughts at Pittsburgh, Rolston once again found himself marching to the beat of a different drum from those around him. At seminary most of his teachers had tried to dissuade him from studying natural theology. Now at Pittsburgh his professors tried to steer him away from the philosophy of nature. They made jokes about Rolston and never quite let go of the idea that he was just a pastor with a quaint interest in the woods. Biology and the philosophy of nature, they insisted, would lead him nowhere. They were too soft and indeterminate. Biological organisms were made up of atoms and molecules. Why would someone want to bother with fuzzy biological explanations when there were more ultimate and determinate physical ones for the same phenomena?

The humor at his expense certainly needled Rolston. More important, though, their reasoning failed to impress him. He had learned firsthand how nature constructed things, not only physically and mechanically but also biologically and historically. Here his Protestant background helped influence his thinking. Transferring an idea from theology to biology, Rolston claimed that biological organisms refine themselves over evolutionary time through their mistakes, *semper reformanda,* always reforming, just as the Protestant theologians had committed themselves to doing. He began to suspect that the philosophy of science needed to take a more naturalistic and historicist turn in order to understand the world. With his master's degree from Pittsburgh moving quickly toward completion, Rolston was certainly better informed about what he needed to do. Perhaps he was also more confident about some of the threads he would use to weave the explanatory fabric together. But with a clearer sense now of where the philosophy establishment stood on some of the issues, he was even more aware of the immensity of his task.

What happened next reinforced Rolston's conviction that the affairs of the world were greatly dependent upon chance. It was a cold day in December of 1967 and Rolston had hurried from the library at the University of Pittsburgh to catch the bus home. He arrived at the stop just in time to see the 5:45 p.m. bus disappearing down Forbes Avenue in a cloud of diesel fumes. Resigned to the half hour's wait, Rolston had just buttoned up his overcoat and turned up the collar when one of his professors, Richard Gale, walked by.

"Rolston," barked Gale. "Are you going to Boston?"

"What's Boston?" Rolston replied.

"The American Philosophical Association annual meeting is in Boston this year. You ought to go. They will be interviewing for jobs." And Gale walked off.

Rolston wasn't sure what to make of the brief encounter. Up to that point he had no intention of going to Boston; he hardly knew what went on at a philosophy conference. He certainly did not anticipate needing to interview at any philosophy job fair. He was already talking on the phone with a couple of schools about returning to the Carolinas or Virginia for a teaching position. Why would he head even further north into Yankee territory for a philosophy meeting? Then he thought a bit more about Gale's suggestion. He wondered what he had to lose by going to Boston. His musings about contingency and chance made him rather intrigued that he had run into Gale after missing the bus. He decided to go to the meeting.

At the conference Rolston had a number of job interviews, nearly all for schools on the East Coast and in the South. He thought he was doing a reasonable job and began to think Gale had been right to urge him to attend. On the last day of the conference, he saw a handwritten note pinned on the bulletin board about a philosophy position at Colorado State University. Rolston had never heard of Colorado State University. Probably some cow college in the West, was his first thought. No need to interview there. Half an hour later he reconsidered, figuring there was little to lose from another interview and he would at least get a better sense of the kind of people philosophy departments were searching for. So he decided to interview with the chair of Colorado State's philosophy department, Willard O. Eddy.

When he met Professor Eddy for an interview later that day, it turned out Colorado State no longer had a job available. The hiring plans in place when Eddy had booked his trip to the conference had fallen through. The two of them had a genial visit. Eddy, it turned out, was a Christian and enthusiastic about the philosophy of religion. They parted on good terms.

Rolston did not hear anything else from Colorado for four months. He had almost forgotten about the interview when late in April he got a call from Eddy. The conversation had barely started when Eddy got right to the point:

"Rolston," said Eddy. "One of our people has suddenly left for California. Do you want a job?"

Rolston was taken completely by surprise. He told Eddy he would need a day or two to think about it.

That night he lay in bed and thought about the prospect of a philosophy job in Colorado. He was more than a little startled at Eddy's abrupt offer. It did not fit at all with the plans he had in mind. He and Jane had never considered the possibility of living anywhere but in the Southeast. Jane, he smiled to himself, being truly a southern lady, would be a little hesitant to move to the West and become a cowgirl. On the other hand, there were certain things that appealed to him about the possibility of a couple of years in the Rockies. Like Daniel Long, the great-great grandfather who had ridden his horse from Philadelphia to Texas, Rolston wanted to see the country. He had a high sense of adventure and knew there would be plenty of that to be found in Colorado.

He had recently made a number of rewarding trips out west. There was a four-hundred-mile raft trip he had taken on the Colorado River through the Grand Canyon. In the piece he had published in the *Bristol Herald Courier*, he had enthused about adventure: "Our water-pony bucked and we held on. The waves crashed over the bow and drenched us, and we held on. This was our initiation, and wild and exhilarating baptism into the muddy and violent Colorado."

He remembered being impressed by how the western landscape, drier and less concealed by vegetation, wore its geologic history on its sleeve. He thought back to the towering rock layers staring him in the face as he lay down to sleep by the river each evening:

> Here am I, as ephemeral as a mayfly, in a canyon of canyons that could bury the East's grandest mountain, nine million years old and carved in the most ancient rocks on earth. I stare upward, realize how close I am to the precipitous walls, and hope this is not the night for the loose rocks above to tumble down.

Memories of other recent trips came flooding back. Three years before he had backpacked in New Mexico with scouts at the Philmont Scout Ranch in the Sangre de Cristo mountains. Another year, he and Jane had made a six-week tour of the western national parks together, camping all the way. They had been entranced by the curious formations in Arches National Park and by the majesty of the Grand Teton. On their last night before the long drive back East, they had camped in Rocky Mountain Park only forty miles from Colorado State University with a glorious view of Longs Peak. While the western mountains lacked the storied family history of the Appalachians, there was ample compensation provided by their sheer, rugged bulk.

The fact that the job was in Colorado rather than Wyoming or Idaho added a particular appeal. The state occupied an almost mythical place in Rolston's family lore. It was the territory where his father's father, Holmes Rolston I, had traveled to save the family farm in the 1880s. The idea of living in Colorado made Rolston instantly feel closer to the relative he most admired. He thought back to some of the stories told of the man he had never met.

Rolston's grandfather had been born to a farmer named John Hopkins Rolston in Harrisonburg, Virginia in 1864. John Rolston had the bad luck to raise his family through the Civil War and the tumultuous years of Reconstruction that followed it. As was common in this period, the Rolston farm teetered constantly on the brink of bankruptcy. Rolston's grandfather had a childhood beset by struggle and hard work.

The poverty turned Holmes Rolston I into a tough child. His skill in the boxing ring earned him the nickname "the Harrisonburg slugger" by the age of twenty. Even as a young man, Rolston's grandfather was independent-minded and industrious. When his father died in 1884 and the family farm was on the point of being lost, the twenty-year-old and his brother Frank headed out to Colorado to make some quick money on the cattle ranches. They knew very little about ranching and almost nothing about the West.

The story goes that on the first night on the Mitchell Ranch near Pueblo, Rolston's grandfather was startled awake by the sound of howling outside. Thinking he was about to be attacked by wolves, he loaded his gun and sat for the rest of the night on his bed waiting for a pack of rabid animals to burst through the door. In the morning, the other ranch hands roared with laughter over their coffee and eggs when they informed him the howling the previous night was only coyotes.

Rolston's grandfather took to ranch life easily. He was strong, with a natural affinity for horses, skilled as a rider and good with a lasso. He also turned out to be a gifted trainer. One of his jobs on the ranch was to "break" horses for riding with the cattle. Rolston's grandfather did not like the idea that you had to destroy a horse's independence in order to train it. He called his work the "gentling" of horses and there was no doubt he was good at it, gaining a reputation around Pueblo for succeeding with horses others had never managed to ride. He quickly gained the respect of his bosses and the other ranch hands.

The young brothers both became competent outdoorsmen, living rough and working long days. Rolston's grandfather gained a passion for fly-fishing. The same observation skills that allowed him to interpret the subtleties of horse behavior also made him good at reading a high country stream. He could anticipate the eddies and swirls that held the big trout. He watched carefully how the stone flies moved above the current.

For just over two years the brothers lived an exciting life in the Colorado rangeland, sending the money they made back home to the family. At the start of the third year, Rolston's grandfather and great-uncle Frank were offered a chance to buy the ranch. In the back of their minds, they had been toying with the idea of moving the family out West. Colorado was a new frontier. Beneath its endless blue skies and soaring snowcapped peaks, the brothers had almost forgotten the unemployment and economic desperation back home. However, there was a rough edge to life in Colorado that held them back. Rolston's grandfather knew the lifestyle would be difficult for a family that aspired to live in genteel Virginia towns like Staunton and Lexington. For this reason, he had never let the life of a frontiersman get too deeply into his soul. With a good chunk of money in hand, the brothers returned home on the train at the end of that summer. They brought back with them a number of western cattle horses to breed with some of the Shenandoah Valley stock, thinking they could create strong eastern horses with the endurance of the Rocky Mountain bloodlines. Back home they quickly worked out a deal to settle the farm debt for fifty cents on the dollar.

Rolston had always trusted the memory of his ancestors to show him the way in life. With his degree from Pittsburgh in hand, he decided it might not be such a bad idea to live in Colorado for a couple of years to broaden his horizons. It would be geographically stimulating and he would no doubt learn a thing or two. Virginia certainly wasn't going anywhere in the short time he planned to be away.

When he mentioned Eddy's offer to his parents, he got dramatically different responses. His father told him to go for it. Rolston's father was always ready to support his son's decisions and thought that a couple of years out West might be good for him.

His mother thought he had lost his mind.

"Why would you want to go do a thing like that?" she exclaimed in horror.

She demanded to know why any young son of Virginia would want to give up the South to go out West, even if was only for a short while. Rolston's mother did her best to persuade him to hold out for a job in a southern Christian school. Jane was also a little nervous about leaving Virginia. The Southeast was her home and the year in Pittsburgh had only made her miss it more. At the same time, Jane was intensely loyal to her husband and possessed a deep desire to see him succeed.

In August of 1968, after completing his master's degree, Rolston and his wife packed up their possessions in Pittsburgh and got ready to move to Fort Collins. One of the family mementos he packed with him was his grandfather's pistol belt, still with some bullets in it, determined to take it back to Colorado where it originated. He knew this was a little bit tongue-in-cheek. There was a big difference between being a cowboy and a college professor.

In August they drove west. As they crossed the hundredth meridian and the land got drier, both of them felt some apprehension. Jane remembers crossing the Wyoming–Colorado border heading south, looking at the arid badlands and razorbacks, feeling a knot tighten in her stomach and wondering what on earth they had gotten themselves into. Rolston himself tried hard not to display any unease. Once he made his decisions, he was not a man to second-guess them. He attempted to reassure Jane that they would be staying there only for three or four years before going back East where—Rolston then thought—the real intellectual action lay. With his fingers gripped firmly around the steering wheel, he fixed his gaze on the road ahead. He knew the next few years would be interesting, but he had no idea at the time how much they would change his life.

On their arrival in Fort Collins, the two Virginians moved straight into a duplex on Baker Street near the university. Feeling highly displaced from their native landscape, they set about adjusting themselves to life in an unfamiliar environment. There were classes to prepare, furniture to buy, and friends to meet.

In just a few short weeks, the crisp nights of early fall would start to turn the aspen leaves in the high country golden.

Portrait of Holmes Rolston I (1864–1924) taken in Colorado, where he worked as a cattle rancher in the 1880s.

Rolston's grandfather, Will L. Long (1864–1951), with Rolston's mother, Mary (left), and her twin sister, Willie Lee, on the Alabama farm c. 1910.

The house across from Bethesda Presbyterian church where Rolston lived for the first decade of his life.

Bethesda Presbyterian outside of Rockbridge Baths, Virginia, where Rolston's father, Holmes Rolston II, preached from 1929 to 1942.

Rolston with his sister Julia by the cistern at the Rockbridge Baths home.

The view of Jump and Hogback mountains near Rolston's childhood home, where his father introduced him to the wonders of the natural world.

A family portrait taken in 1942 shortly after Rolston's father transferred from the Shenandoah Valley to a new church in Charlotte.

Rolston with Jane Wilson at a Davidson College dance in the early 1950s.

On board the HMS *Queen Elizabeth* en route to the University of Edinburgh in 1953 to study for a doctorate in theology.

Rolston in his first job as a pastor at High Point Presbyterian in 1961, before being asked to leave for being too out of touch with his congregation.

Rafting the Grand Canyon in summer 1967.

Rolston in 1972, early in his teaching career at Colorado State University, with his children Shonny (left) and Giles.

Reunited with his former advisor and Templeton Prize winner Tom Torrance during the Gifford Lectures in 1997.

Receiving the Templeton Prize from His Royal Highness the Duke of Edinburgh in 2003 at Buckingham Palace in London.

Distinguished University Professor at Colorado State University where Rolston taught from 1968 to 2008.

Hiking the Selway Craggs in Idaho, age seventy-three.

PART III

Rocky Mountain Philosopher

7 Colorado Breakthrough

Driving across Wyoming or Colorado, there are those ghost towns, or abandoned ranches, reminders of failed settlements. The old farmhouse, a ruin, again with failing fences, invites us to reminisce about the past. But the cabin is falling into ruin because it is incessantly weatherbeaten by an ever-present returning nature. There is a sense of the transience of human endeavors. The struggle for survival lurks in the landscape. [*2008*]

ROLSTON WASTED LITTLE TIME before exploring the geography of his new home. Colorado State University was located in the Front Range of the Rocky Mountains where the foothills met the plains. To the east, the land lay flat and uninterrupted all the way to Kansas. On the western edge of town, red sandstone plates jutted abruptly from gently contoured rises peppered with sagebrush and mountain mahogany shrubs. The sandstone edges were a humble forewarning of the geologic dramas in the stratigraphic column behind. Five miles west of Fort Collins, the ponderosa pines of the foothills gave way to lodgepole pine and subalpine fir as the country tipped abruptly higher toward the mountains and alpine tundra.

The Cache la Poudre River, named to commemorate a group of French trappers who buried their gunpowder on its banks during an 1820s snowstorm, wound its way east toward Fort Collins from its headwaters in the glaciated peaks of the Never Summer Range. Forty miles to the south and west, Rocky Mountain National Park, the Indian Peaks, the Rawah, and the Comanche Peak wilderness areas were host to more acres of alpine ecosystem than the whole of Switzerland. The area around Estes Park was home to one of the densest populations of elk in North America. Amid the towering peaks and endless forests of Colorado, Rolston began the process of grounding himself in his new environment, turning his naturalist's eye to the unfamiliar local flora and fauna that he explored at every opportunity on foot, horseback, and snowshoes. While adjusting to life in Colorado, he made his first entries in a new set of trail logs.

> *January 18, 1969.* Rocky Mountain National Park. Spent the day deciphering the peaks. Late afternoon saw 32 mule deer, few or no bucks, and 61 elk, many bulls. 24 elk were in Little Horseshoe Park, and the remainder in Horseshoe Park. Some majestic bulls.
>
> *January 25.* Drove to Ben Delatour Scout Camp and hiked in that area. Broken woods, mesas. Ponderosa pine, sage. Some sign of elk, rabbits. Light snow cover. Nice view of perched hawk, later soaring. Probably a rough-legged hawk.
>
> *February 1.* Hiked up Hewlett Gulch, from Poudre Park, to Bidwell Mining Claim. Gentle to negligible climb. Sage-Ponderosa belt. One adit open and easily accessible. The mine seems to be worked enough to keep the claim open.

When Rolston speculated at the philosophy meeting in Boston that CSU was probably some cow college out West, he was not altogether wrong. Fort Collins itself was established as a military garrison in 1864 and incorporated as a town shortly thereafter. Within a decade, the Colorado Central railroad had arrived and the first buildings of a new college had been erected. An eight-person Colorado State Board of Agriculture was appointed to run the new school. In 1879, the first five students enrolled at the freshly named Agricultural College of Colorado.

The college was one of what would eventually be seventy land-grant schools in the United States owing their existence to the 1862 Morrill Act. To support higher education, the act granted to every state in the union thirty thousand acres of federal land for each member of their congressional delegation. The language of the act stipulated the use of funds:

> [To establish] at least one college where the leading object shall be, without excluding other scientific and classical studies . . . to teach such branches of learning as are related to agriculture and mechanic arts, in such manner as the legislatures of the State may respectively prescribe, in order to promote the liberal and practical education of the industrial classes in the several pursuits and professions in life.

Particularly in the middle of the country and farther west, the Morrill Act provided an important boost to university education at the end of the Civil War. Each state gained at least one college with programs centering on agriculture and natural resource sciences.

By the time Rolston arrived to teach at Colorado State in the fall of 1968, the college had evolved through three name changes and eight presidents, but the original mandate as a land-grant institution remained a significant part of the school's identity. Its Departments of Forestry, Agriculture and Resource Economics, Animal Science, and Range Management were world class. Other departments, with names such as the Departments of Bioagricultural Sciences and Pest Management, Soil and Crop Sciences, Geosciences, and Fishery and Wildlife Biology, all indicated the university was still true to its land-grant mandate. Its Natural Resource Recreation and Tourism Department provided the majority of employees for the Colorado State Parks system. CSU graduates were district rangers, forest supervisors, and park superintendents all over the West. Biology-related classes were taught in more than two dozen departments. Rolston realized he had traveled a long way from the rarified scholastic atmosphere of the East Coast Christian colleges to which he had aspired. The world-class philosophy of science department at the University of Pittsburgh also felt like it had receded far behind in his rearview mirror.

As a backdrop to all of the university's academic activity, the Rocky Mountains sat impressively on the western horizon, visible from almost every building on campus. Reaching more than fourteen thousand feet into the endless blue Colorado sky, the soaring summit of Longs Peak presided over the campus below. Rolston had cherished the mountains back home in Appalachia. Here in Colorado, the snow-capped peaks created a mental ambience of a quite different order. It was a perfect place for a philosopher of nature to take root.

Immediately, Rolston found the physical environment stirring thoughts about the human place in the world. Western Colorado did not let its inhabitants escape the sense of geologic history projected by its mountains. Having moved from the more intimate landscapes of the East, he was taken aback by how forcefully the Rockies inserted themselves into the people's consciousness. These mountains were geologic giants, starting their steep rise from the elevation at which Rolston's familiar Appalachians topped out. Much later, Rolston wrote that Coloradans think of the mountains as "a defining part of our existence, palpably affecting our sense of presence. . . . The great vista of life on earth is still evident on the horizon of the local landscape." Both nature and culture tugged hard at one's identity in the Colorado high country. This thinking began to feed into the development of ideas he was now, for the first time in his life, finally free to develop.

A reconciliation of Christian theology and the biological sciences remained Rolston's primary goal, but he was by now feeling compelled to spend more of his time arguing for the obligation to protect nature. Clearly the two projects could be connected. A well-worked-out theology of nature would likely include a serious moral obligation to the natural world, bringing together the two intellectual currents in his life. On his entrance into the philosophical community at Colorado State, the connection between the two research projects quickly became far more complicated than he had ever imagined.

When Willard Eddy had extended his offer to join the CSU faculty, there had been some quiet murmurings of discontent among a few of those already on the faculty. While several of the professors had their own interests in comparative religion, others were immediately suspicious about hiring a pastor. Philosophers see it as their job to promote the critical use of reason in their students. Why, then, were they hiring a religious man, a Presbyterian pastor no less, without a doctorate in philosophy, more used to preaching than teaching, to lecture in their classrooms? How could they succeed in encouraging their students to think for themselves when these same young men and women would sit in the next class listening to a man who based some of his most fundamental commitments on faith? He seemed like a polite and serious individual—if a little reserved—but his colleagues wondered, were Rolston's skills in philosophical argument going to stack up?

In private life, he and Jane were already making friends at the First Presbyterian Church of Fort Collins on College Avenue. At church they could wear their religious commitments on their sleeves. But in his professional life, Rolston knew he would have to keep his personal beliefs off the table. He would need to firmly establish his philosophical credentials if he was to win his colleagues' respect. In terms of his academic research, this meant keeping the theology in the background. He needed a strictly philosophical research program even when he knew in his heart that questions about the divine were hovering in the background. Jane could tell from her husband's behavior that he keenly felt this pressure from his peers.

In his quest to clothe the theology in philosophical attire, Rolston found plenty of inspiration within philosophy's own history. The interest in writing secular justifications for what were essentially theological positions had been common among philosophers from the beginning. Toward the end of the eighteenth century, the devout Protestant philosopher Immanuel Kant had recognized the necessity of nontheological arguments in moral philosophy. A central problem for Kant was how to formulate an ethic universal in its scope, applying to every human on earth. He realized that philosophy should provide a system of morality valid for believers and nonbelievers alike.

Both Rolston and Kant were confronting an ancient philosophical problem. Like many other philosophical puzzles, it had been anticipated by Plato more than two thousand years earlier. In a dialogue Plato entitled the *Euthyphro*, his protagonist Socrates was discussing the topic of human goodness or piety. In response to Socrates' question, Euthyphro had defined piety as "what all the gods love." Socrates pointed out the conundrum contained in his remark: "Is something pious because the gods love it, or do the gods love it because it is pious?"

Euthyphro was forced to admit that piety had to be more than a random matter of taste and preference, even if the preference being considered was divine. There had to be a reason why the gods considered something pious before they could love it. But what was that reason? Socrates quickly convinced Euthyphro that he did not yet understand the nature of piety.

Kant recognized the problem presented in the *Euthyphro* and saw that the solution lay in finding a rational argument for morally acceptable behavior that also coincided with the commands of his faith. This meant a secular argument for his religious ethics. Or, to put it in Kant's more technical language, a rational reconstruction of a theological position. This approach would satisfy Kant's colleagues in the philosophy department at the university in Königsberg because he would rely on reason rather than faith to make his case. It would also satisfy his colleagues in the theology department because he would recommend moral laws they had already embraced for religious reasons. The arguments Kant went on to articulate about the existence of a "categorical imperative" in ethics made him into one of the most influential thinkers in the history of Western philosophy.

Whether or not Rolston had Kant in mind when he started work on becoming a secular philosopher at Colorado State is unclear. Rolston knew that to satisfy his philosophy colleagues, to say nothing of the foresters and biologists on campus, he would have to create an ethic of nature free of religious grounding. This was true even if he ended up advocating a position that coincided, as Kant's had done, with his Christian beliefs. Rolston accordingly drew a mental line in the sand and determined, at least initially, to keep his religious commitments out of the classroom and the department hallways. While this was something of an accommodation to the politics of the situation he found himself in, from a career perspective, it turned out to be a very fortunate choice indeed.

As he pondered how to develop some of his thoughts on nature's value, Rolston asked himself if he did not already have a secret weapon at hand. He could use his knowledge of the natural sciences to formulate an ethic that was reason-based and therefore nonreligious in the eyes of his philosophy colleagues. It was well known that science and religion resided at the opposite ends of the intellectual spectrum. Science was based on observation and logic, religion was based on faith and revelation. Thanks to the classes he had taken at East Tennessee State and the philosophy of science he had studied at Pittsburgh, Rolston felt with some justification that he knew the scientific methodology quite well. If he could create an ethic rooted firmly in the natural sciences, he might dispel the worry about his philosophy being too religiously motivated.

Quickly he ran into a problem. Almost all philosophers insisted that ethical philosophy be kept separate not only from religion, but also from natural science. Science, after all, was designed to provide just the facts. Scientists who wanted to keep their jobs carefully avoided making moral assertions. Ethical philosophy, by contrast, placed moral overlays on top of the facts. Earthquakes, drought, and disease were facts about the world studied by scientists. Care for the homeless, compassion for the hungry, and sympathy for the families of those killed by epidemics were moral sentiments designed to guide people in light of those facts. Scientists had traditionally been banished from these ethical arenas. To attempt to mix the two inquiries was to commit a serious philosophical error. Facts and values belonged in two separate domains, each with its own experts.

At the time Rolston starting reflecting on these issues, the need to keep the science and the ethics apart had assumed increasing urgency due to a developing turf war. Philosophers had always insisted that they alone were qualified to discuss matters of right and wrong. Starting in the late 1960s, a small group of scientists, inspired by the "survival-of-the-fittest" language in evolutionary biology, had begun to expand their domain to cover explanations of human behavior. Since all biological life had the appearance of being self-interested from the evolutionary standpoint, human social life, these scientists presumed, also reflected this universal phenomenon. The new science of sociobiology being promoted by Harvard biologist E. O. Wilson developed around the claim that all apparently altruistic behavior was in fact a cover for genetic self-interest. People were biologically driven to promote the survival of their offspring. Charity came with the unspoken expectation of something coming back in return.

Philosophers were infuriated at this presumptive grab of territory by evolutionary biologists—"biological imperialism!" they called it. Ethicists saw it as their duty to act as gatekeepers against the intrusion. Unfortunately for Rolston, the strategy of promoting his scientific credentials to reassure his colleagues looked like it was sure to get him into trouble. An ethic with its starting point in science, no less than an ethic with its starting point in religion, would face extraordinary barriers against acceptance in the philosophical community.

Rolston now found himself caught at the intersection of three disciplines, not two, pulled in a different direction by each. Science, religion, and philosophical ethics were unlikely collaborators. Yet it was precisely at their intersection that Rolston wanted to work. He thought of himself as being caught up in a lovers' quarrel, difficult and painful to resolve for sure, but rooted in what had to be an intimate connection. If he was ever going to develop a philosophy of nature, he would need to figure out how to connect these three disputants in a way that left all of the disciplinary prejudices, as well as the egos of all those involved, intact. The former Virginia pastor, now crafting an identity for himself as a philosophy professor at a large state university, had difficult work ahead of him.

Whenever he felt lost or overwhelmed by the challenges in his new academic environment, Rolston followed a familiar tactic. He only had to step out of his front door to find continued inspiration for his intellectual quests. Rolston was drawn to the numerous wild areas easily accessible from Fort Collins. He invested in a full set of topographic maps covering a hundred miles in each direction and started methodically hiking the trails. While other new professors were caught up in overpreparing their classes or fretting about how they were being received by their colleagues, Rolston was spending most of his weekends with boots on his feet, field guides in his pack, and a hand lens at the ready. Some of his colleagues wondered if he was shirking his professional obligations. For Rolston, discovering the names of the trees and shrubs, learning the geology underneath his feet, and reading the tracks of animals in the snow was not wasting time or avoiding work. It was a way of doing philosophy. The land, quite literally, was feeding his mind.

Always in the shadow of Longs Peak, Rolston toiled away at his intellectual project. A field trip with Mac McCallum, a geology professor from Colorado State, got him thinking about dirt. His trail log documented the geologic richness:

> *May 28, 1969.* Field trip. Over Horsetooth, descending through the geological sequence. Dakota, Morrison (a gray, weaker rock just past summit going westward; the famous dinosaur formation), then into a Lykins Valley. The Lyons is next; a ridge maker, famous for building stone; CSU is built of it. Next is Satanka, a little valley maker, then Ingleside, a ridge maker. The Fountain follows, a big valley maker. The Fountain at Boulder and Colorado Springs form the Flatirons and Garden of Gods respectively.

Pitting himself against those who thought them useful only for logging and mining, Rolston wanted to show that these Colorado mountains themselves were of moral significance. Put bluntly, he had to persuade people to take a moral interest in dirt. He had not yet read Aldo Leopold's *Sand County Almanac*, but he needed an ethic directed not simply at the well-being of people but also, as Leopold had famously recognized, at the land itself.

The suggestion that mere dirt could have moral value at first looked like the ultimate absurdity. Philosophers he encountered at professional meetings made jokes to him about the claim. The word "dirt" was almost a synonym for the word "undesirable." Nobody doubted that dirt had some value when used to serve a human purpose, for example, to build the foundation of a house or to create a flower bed in the backyard. The idea that dirt could be valuable aside from any human use—"in-itself" or "intrinsically" as philosophers liked to say—was close to a contradiction under the common meaning of the term.

Rolston felt that if you correctly read the mountains they had something different to say about dirt. The western landscapes gave him a whole new appreciation for the value of dirt. On his field trips with McCallum, Rolston learned the storied history contained within rock and earth, "raising a curtain on a drama already in progress." The layers of history written into the strata and deciphered by the geologist could not fail to impress anyone with even a modicum of curiosity about how we got here. "Pretty spectacular dirt!" Rolston remarked when gazing with McCallum from the top of Horsetooth Mountain to the dinosaur formations at his feet. The rocks were riddled with stories of a remarkable journey through time. He recalled his earlier admiration for rocks back in the Appalachians at Pardee Point, above the Doe River Gorge. The realization when gazing into the gorge that all life originated from ancient rocks had prompted an almost religious reflection at the time in praise of "the lifeless granite, this fossil conglomerate inescapably now become charged with a numinous mystique." In rock, he recognized, lay the human geogenesis. Rock was ultimately transubstantiated into soul, something he knew was also written in his Bible.

Rolston's insight about the importance of historical context gained at seminary now took root in his thinking about the land itself. Neither the natural nor the cultural history of Colorado could ever be fully understood without the stories in its dirt. Every handful of earth connected to a complex sequence of geologic and hydrologic events. These were not just the stories of early pioneers and settlers, but stories of geologic upheaval and destruction, of trilobites and saber-toothed cats evolving and perishing over extended sequences of time. In addition to stories of cataclysmic destruction, the rocks also told stories of earth's magnificent creativity.

Dirt, Rolston knew, was only the beginning of natural history. Things became even more impressive when one added the biological complexity of the organic matter found on top of the bedrock. The tales told there were little short of astonishing. He read all he could about evolutionary biology in the CSU library and found many biologists similarly impressed. E. O. Wilson detailed the complexity best:

> Think of scooping up a handful of soil and leaf litter and placing it on a white cloth—as a field biologist would do—for closer examination. This unprepossessing lump contains more order and richness of structure, and particularity of history than the entire surface of all the other (lifeless) planets. It is a miniature wilderness that would take almost forever to explore. . . . Every species living there is the product of millions of years of history, having evolved under the harshest conditions of competition and survival. Each organism is the repository of an immense amount of genetic information.

Literally millions of creatures each with millennia of history occupied every handful of organic matter. A single cubic inch spilled over with countless tales of life evolving and being extinguished. Impressive above all else to Rolston was the fact that dirt contained the ever-present promise of new life. If there was an appropriate substrate for ethical value, a system of rocks four and a half billion years old with the facility to generate and support life must surely be it. Rolston saw the earth's landscapes as miraculously fertile ground. *De Profundis* indeed! His secular philosophy of nature was getting its grounding in dirt.

Unfortunately for Rolston, these intuitions soon crashed into the barriers placed before him by his academic colleagues. The logical positivists in his department and the hard-nosed scientists across campus were unwilling to make the move from the impressive geology and ecology to Rolston's claims about moral significance. Even if natural history revealed a world of immense geologic and biological complexity and even if the complexity was in some sense impressive and even awe-inspiring, most philosophers insisted this had no bearing whatsoever on matters of moral value. Science, they maintained, just reported the facts. Moral claims resided in an entirely separate realm. As his colleagues had predicted, the ex-pastor was sermonizing, not philosophizing. To suggest something possessed moral value simply because it was scientifically interesting was to make the forbidden crossover from science to ethics, facts to values. Rolston knew he disagreed. He just did not yet know how to argue it.

Temporarily at a loss for how to get from the natural science to the moral value, Rolston's mind drifted back to Davidson. By showing him the unnamed insect, Buggy Daggy had prompted Rolston's conviction that biology rather than physics was the most explanatory science. Was it possible that somewhere within biology lay the key to a workable account of the moral value of nature? Biology, after all, studied life, and life seemed to count for an awful lot in ethics. Every theological and ethical bone in his body told him that the study of life was linked to the study of value. Rolston wondered whether, by sticking close to the biology, he might find some solid connection between the moral values present in the human world and the moral value he glimpsed within the ecological systems from which humans had emerged.

The Colorado landscape played a significant role propelling him forward in his quest. On every hike in the high country, the power of biological survival in the face of harsh environmental forces called out its significance. At the start of his first Colorado spring, Rolston had been taken aback on a still, cold hillside in the Rawah Wilderness when he came upon the purple blossoms of thousands of pasqueflowers reaching skyward toward the sun's warming rays. He reflected on the meaning of the pasqueflower's blossoms. To push its flowers up through patches of melting snow at the beginning of a new season of growth, this delicate flower seemed to indicate something about life that reached far beyond any botanical observation. The pasqueflower, Rolston wrote,

> helps us to celebrate because it dares to bloom when the winter of which we have wearied is not yet gone. "Flowering" touches values

so soon; this biological phenomenon becomes a metaphor for all the striving toward fruition that characterizes the psychological, intellectual, cultural, and even the spiritual levels of life.

This striving toward fruition in the pasqueflower's vernal florescence appeared to mirror something about the striving present in numerous human struggles in pursuit of something better. The flower's tenacity under difficult conditions reminded him of his grandfather's father heading west to earn money when the family farm faced bankruptcy back home. It made him think of his mother's great-grandfather becoming a physician after the 1812 war and of her father fighting against hunger and poverty in Depression-era Alabama. Something of value had accrued through the hard-won successes of these forebears. Despite the suffering and sometimes the perishing, life persisted, its essential core enduring. Descendants had been inspired. Character had been built. In both the pasqueflower's and the ancestors' lives, the challenges presented by the darkness of winter made possible the creative potential of spring. Somewhere within this common truth, Rolston was starting to believe, lay an argument for nature's deeper values.

Rolston could already hear his colleagues' objections ringing in his ears. Botanical nature differed radically from human culture, they would insist. The common struggle he proposed was merely metaphorical. His praise for the pasqueflower was just romantic poetry. There were moral values in the human world only because humans created a shared vision of community in which those values were rooted. Those shared visions were embodied in social institutions, customs, and laws. They were products of the arts, of community, and of culture. In each of these arenas, humans were radically different from any of the organisms in the natural world. Wildflowers blooming in spring staked no claims about right or wrong. There was nothing moral in nature, just the simple facts of biological survival. To make a solid link between the bloom of the pasqueflower and the moral values handed down by his ancestors, Rolston knew he needed something more concrete. The connection between culture and nature must become less metaphorical and more literal, and have fewer rhetorical flourishes and more objective reports about plants and animals defending values in their lives. Looking at the etymology of the word "biology," Rolston asked himself again what exactly was meant by a word that translated as the "logic of life."

Since his arrival at Colorado State, Rolston had continued the habit developed in Virginia and Tennessee of sitting in on classes in the natural sciences. The link between biological survival in nature and the moral lives of humans came to Rolston one morning when listening to a lecture on evolutionary biology. The professor made a comment about Darwin's work being insufficient for a full understanding of evolution. He claimed that evolutionary theory only became persuasive when Darwin's theory of natural selection was combined with Mendelian genetics, a synthesis known as "neo-Darwinism." Genetics turned a rather speculative Darwinian proposal into a full-blown theory.

Rolston had known this tidbit in the history of evolutionary biology for years, but this time the comment took on particular significance. The genetics, Rolston realized, did add something significant to Darwin. It added the critical phenomenon of the passage of information from one individual to the next. Genetics gave Darwinism the key mechanism it needed. Genes made it possible to pass important information on to a new generation. Information transfer was essential to the logic of life. It also gave Rolston the insight he had been waiting for.

No philosopher would deny that the passage of information between humans was one of the crucial factors at the center of culture and morality. There was no possibility of morality without information transfer. Humans passed on information through numerous mechanisms, including books, universities, museums, parental instruction, and religious traditions. Moral values hinged almost entirely upon the possibility of this information transfer. Wisdom about right and wrong could not be transferred without it.

Rolston now saw that biological organisms also passed on information to their descendants, not consciously as humans did, but biologically through the codes written into their DNA. When the pasqueflower spread its petals in spring, leading eventually to the production of seeds, it was getting ready to pass on the information required by the next generation of pasqueflowers. Like humans, pasqueflowers needed their ancestors' know-how to survive and flourish. DNA enabled the future generations to learn from the trials of those that came before. Information transfer was the ultimate ground of all achievements in both culture and nature. It was the crucial phenomenon connecting biology to morality.

To make the connection clearer, Rolston started calling DNA a *linguistic* molecule, one that effectively provided biological life with a storied memory of the past. "Humans artificially impose an alphabet on ink and paper, but living things long before were employing a natural alphabet, imposing a code on four nucleotide bases strung as cross-links on a double helix." DNA, essentially an information-bearing molecule, was a stroke of biological genius. Genes enabled an organism to inherit lessons accumulated over the entire history of a species about how to live and grow. A genetic set was a memory of how to triumph over difficult conditions. Information transfer via genes was the conceptual link between the physical chemistry of protein chains and the institutions at the heart of human culture. It was the connection between what physically *is* in nature and what morally *ought to be.*

The argument Rolston now had at his fingertips was the culmination of years of personal experience in nature. Rolston retreated for a week alone at Lake Solitude, remote in a Rocky Mountain park, forcing his mind through the logic he had stitched together. The encounter with the lake's wildness nourished the dialogue between natural history and philosophy taking place in his mind. "I may come alone," he reflected, "but I understand with the genius of a multitude of minds, all multiplying and amplifying my solitary enjoyment." In front of his mind was the cultural wisdom. In front of his eyes were billions of years of encoded biological know-how. Images and memories of earth's numerous achievements cascaded forth. The purple petals of the pasqueflower against a late spring snow, the defensive clutch of partridges holding their ground in Virginia, the fox darting into the hedgerow and out of range of the farmer's gun, all demonstrating generations of wisdom handed down from the struggles of ancestors. The biological inheritability of information was the ground of all achievement and ultimately the ground of all natural value. With DNA, Rolston had found the key to the connection between moral values in the human world and moral values in the biological world. He was impatient to try these ideas out on the philosophical community.

His peers, it turned out, were skeptical. From the perspective of traditional philosophy, these thoughts were a little too wild. The disciplinary prejudices were stacked completely against the position. Rolston's first attempts to publish his theory of an ethic of nature based on DNA as a "linguistic molecule" met with rejections and disappointment. In 1974, with his frustration leading to a little impetuousness, he submitted an article spelling out these thoughts to the philosophy journal *Ethics.* As the world's leading journal in moral, legal, and political philosophy, it had been publishing cutting-edge work for close to a hundred years. If a person could get an article published in *Ethics,* everyone in the field understood, the philosophy must be first-rate. Rolston felt he had little to lose.

To his great surprise, *Ethics* accepted the article almost by return mail, publishing it in January of 1975. The article, titled "Is There an Ecological Ethic?" is still considered to be not only the breakthrough article in the career of Holmes Rolston III but also the breakthrough article in the field of modern environmental philosophy. With its publication, the idea that natural systems (and the organisms they contained) were morally significant immediately assumed philosophical respectability. On the back of this one powerful idea, the field of professional environmental philosophy had been born.

8

Valuing Wild Nature

> Modern man came out of the Ice Age. . . . We do not owe every culture to the Pleistocene winter, for archaic civilizations arose in the tropics, but we owe all culture to the hostility of nature, provided only that we can keep in tension with this the support of nature that is truer still, the one the warp, the other the woof, in the weaving of what we have become. [*1986*]

VERY QUICKLY AFTER PUBLISHING his breakthrough article, Rolston started cementing his place at the vanguard of the emerging discipline of environmental ethics. Initially, a good deal of the work done was to persuade a rather skeptical philosophical community there was a viable case to be made for the existence of values inherent in nature. A number of his colleagues at Colorado State were astonished that Rolston's unorthodox ideas about natural value were getting any attention at all. Part of the reason for the initial traction was that a handful of other philosophers had already done crucial work preparing some of the ground. An Australian named Richard Routley had published a paper in 1973 titled "Is There a Need for a New, an Environmental Ethic?" outlining what a true environmental ethic might look like. A Norwegian thinker and alpinist, Arne Naess, published an article in the same year titled "The Shallow, and the Deep, Long Range Ecology Movement," distinguishing human from environment-centered ethics. In both of these papers, the authors had suggested that nature possessed moral significance independent of human uses or interests.

Up to this point in philosophy, nature's only value had been routinely assumed to be as an instrument for the satisfaction of human needs. Trees, for example, were useful instruments for building houses or for crafting furniture. Clean streams were useful for recreational fishing and irrigation. Wildlife species provided benefits for hunters and photographers. Each of these instrumental approaches to nature's value stopped short of considering the moral standing of nature independent of its human uses. Rolston, Routley, and Naess together challenged this way of looking at the world. They asked whether something important was being missed if one assumed that value in nature always had to have some human reference. Could nature, independent of human interests and desires, have some value in itself or *intrinsically,* these philosophers asked?

The argument for the presence of intrinsic value in nature was radical because it challenged a two-millennia-old assumption about the proper scope of ethics. Almost all ethicists up to this point had quite reasonably assumed that their central concern was to debate the principles underlying the proper treatment of humans by humans. The assumption was that the slow transition through history from conditions of barbarism to those of civilized society had depended upon the acknowledgment that there was a proper way for people to treat each other. Ethics was a people-to-people affair and philosophy was in the business of discovering the what's and the why's of this proper treatment. Only a few figures in philosophy's history, most notably the utilitarians Jeremy Bentham in the eighteenth century and John Stuart Mill in the nineteenth, had ever been willing to consider there might also be rules for the treatment of nonhumans. Despite the fact that several national animal welfare movements appeared in the mid-nineteenth century, Bentham's and Mill's early gestures to expand ethics toward nonhuman nature were generally ignored by professional philosophers.

It is perhaps unsurprising that ethics started out human-focused, but one of the effects of this focus was that ethics slipped from being *primarily* about humans to being *exclusively* about humans. Instead of asserting simply that humans were the most important concern in ethics, philosophers in the Western tradition, supported by a mixture of the culture's Christian roots and its powerful humanistic institutions, insisted that humans were the one and only concern. Adding to these historical factors was the unfortunate psychological tendency to think that "might makes right," the belief that those with the power to enforce the rules also had the authority to make them. "Might makes right" left animals and plants repeatedly at the whim of earth's most powerful species. The early work in environmental philosophy by Rolston, Routley, and Naess laid down a direct challenge to these ancient assumptions.

Almost coincidentally with Rolston's breakthrough work in environmental philosophy, a young Australian philosopher named Peter Singer published *Animal Liberation,* a hugely influential book that led to sweeping changes in how society viewed domestic and captive animals. Singer developed an argument, foreshadowed by Bentham and Mill, for why the suffering of nonhuman animals should matter.

According to Singer, the ability to suffer is absolutely fundamental to the concept of right and wrong. Furthermore, suffering matters wherever it is found. To insist that only the suffering of one's own species counts for anything is both chauvinistic and philosophically unjustified. To fail to take into account the suffering of those outside of one's own species exhibited to Singer the same type of prejudicial thinking employed by racists and sexists to discount the interests of others. Singer focused on two particular areas of animal suffering, meat production and animal experimentation. He did not produce a watertight argument against all meat eating and all animal testing, but he argued that the majority of suffering humans caused animals in these two areas served only very trivial human interests and came at great cost to the animals.

While at first glance it looked like Singer and Rolston were pressing a similar progressive agenda, the philosophical differences between their positions soon became apparent. Like Rolston, Singer had crossed an important threshold in ethics by arguing that matters of right or wrong extended beyond humans. After this initial commonality, however, their positions quickly diverged. Singer was concerned only with the suffering of animals, whereas Rolston was more concerned with the health of species and ecosystems. Singer based his ethic on a criterion proven to work well in human ethics, namely the capacity to suffer. Rolston introduced an entirely new criterion, namely the vital information-carrying capacities of DNA and the values accumulated in ongoing species lines. Singer was not particularly interested in wild nature, focusing his attention more on slaughterhouses and medical research facilities. In Rolston's case, wilderness and the health of natural landscapes were absolutely central to ethics. Applying his ecosystem approach, Rolston could support hunting. Focusing on the suffering of individuals, Singer had to oppose it. As a result of these differences, philosophers soon separated Singer's animal welfare approach from Rolston's environmental focus.

Despite this parting of the philosophical ways, the ethical views articulated by Rolston, Singer, Naess, and Routley in the mid-1970s together achieved something dramatic and provocative. For the first time in Western cultural history, ethicists had successfully pressed the case that humans should constrain their behavior for reasons unconnected to their own interests. This new group of progressive philosophers requested limits on human behavior circumscribed by the needs of nonhuman others. Rather than arguing that society should not clearcut a forest because the eroded hillsides would be a threat to those living below, Rolston claimed that the forest had an intrinsic moral integrity that it would be wrong to destroy. Though Rolston's ethic did not discount the importance of human interests, it did not rely on them to justify the proper treatment of nature.

Back in Fort Collins, Rolston was about to celebrate his forty-fourth birthday. Though the argument in "Is There an Ecological Ethic?" had been crystallizing in his mind since he starting wandering the countryside around High Point Presbyterian Church in the late 1950s, to publish the thoughts at this particular moment in history was an uncanny piece of good luck. In the late 1960s and early 1970s, environmental awareness had exploded across the United States. Membership of the Audubon Society had risen from 45,000 members in 1966 to 193,000 in 1972. The Sierra Club had grown from 35,000 to 137,000 in the same period. The United States Congress had added considerably to its initial environmental awakening in the late 1960s by amending the Clean Air and Water Acts, and adding the Marine Mammal Protection Act, the Wild Free-Roaming Horses and Burros Act, the Endangered Species Act, the Coastal Zone Management Act, the Forest and Rangeland Renewable Resources Planning Act, the Toxic Substances Control Act, and the Whale Conservation and Protection Study Act in addition to numerous other laws. More than a dozen national parks, recreation areas, seashores, and preserves had been established. Several million acres of public land had been protected as wilderness—the highest level of federal protection afforded in the United States. When Rolston started to articulate the idea that nature and natural systems had their own intrinsic value, his arguments resonated with a growing cultural awareness of the importance of the natural environment. Increasing numbers of people were starting to see the protection of nature as a moral issue.

The vast majority of the nascent environmental public recognized that clean air and pollution-free rivers would be valuable to their children and their grandchildren. A subset of more progressive citizens, awakened by the writings of Peter Singer, appreciated how the health of the environment mattered not only for people but also for the animals populating various ecosystems. Images of wild animals at play were by this time regularly brought into people's living rooms on television. Charismatic creatures such as gray whales, sea otters, and dolphins held a powerful place in public affection. To increasing numbers of Americans, coastal pollution seemed morally wrong not just because of its impact on humans but also because of its impact on fellow mammals.

Rolston's arguments for natural value pushed the ethical frontier to a whole new level. Natural values did not rely upon the interests of conscious beings, humans or whales, to exist. The values were embedded in geologic and evolutionary history, inhering in earth's dynamic system whether or not there were people—or even whales or sea otters—present to appreciate them. Ultimately, Rolston argued, the dirt itself mattered. Even inanimate nature made moral claims on us. Though this was not a position everyone was ready to embrace, the ideas of this soft-spoken and well-mannered ex-Virginia pastor emerged at a moment in history when they had a tantalizing popular appeal.

Unfortunately for Rolston, what goes down well with the public and what goes down well with professional philosophers can sometimes be greatly different. The small but growing cohort of environmental philosophers now pushing the idea of intrinsic natural value faced continuing derision from their more traditional colleagues in the mainstream. In some cases, the traditionalists did not recognize them as doing philosophy at all. To persuade a still skeptical audience, Rolston used more than just logical argument to articulate his ideas, unpacking his new theory with clever word play. The claim that "nature is intrinsically valuable," he suggested, really amounted to the belief that "nature is *value-able*," that is, "able to generate and support values." He elucidated on the philosophical claim of value support by cataloging different ways of thinking about the moral potential contained within nature.

Some of this potential was realized, as philosophers had recognized for over two thousand years, when humans used nature for their own purposes. Natural systems, for example, supported recreational, economic, aesthetic, and character-building values. These values manifested themselves when people found a particular way to use nature for some benefit. In these cases, nature was employed as an instrument to serve a human purpose. Though he acknowledged the importance of this "instrumental value," Rolston made it a point to emphasize that such values exist only because nature was rich enough and productive enough to be an instrument for satisfying human desires.

Searching for a way to do justice to this productivity, Rolston dug deeper and articulated other natural values not dependent on human desires and interests. First, he noted the wealth of complex and diverse life-forms that had appeared over evolutionary time. Rolston suspected that even the most self-interested human must at some level admire many of these other life-forms. Next he unpacked some of the ecological relationships between them, showing how organisms always depended upon a myriad of other living beings and the web of services they provided. For humans, these services ranged from the bacteria living in our intestines, to the edible plants and food animals that kept us alive, to the fungi that decomposed our waste products, to the phytoplankton in the oceans that absorbed carbon dioxide, to the forests that regulated the hydrologic cycles across the globe. Even a casual look through an ecological lens revealed an earth teeming with interrelated biotic support structures.

Rolston then reasoned that if we valued earth's biota instrumentally for their ability to support us, then it was reasonable to value the simpler lives responsible for the supporting. These lives did not gain their value simply from their use by more important species. The values were intrinsic to the organisms themselves. Only organisms that contained their own value could be used as value sources by others. After all, life was built on DNA. Hence all organisms—bacteria included—contained their own story about an ancestral struggle to survive and pass on information to their descendants. The fish that nourished Rolston during his week of reflection at Lake Solitude and the insects and larvae that nourished the fish all had moral standing. Every living being, as Rolston saw it, from bacteria and viruses to mosses and grizzly bears, possessed this intrinsic natural value.

After instrumental and intrinsic values, Rolston identified an additional layer of natural value. Spread even more widely than either the instrumental or the intrinsic values were those facets of the system responsible for generating both life and life support in the first place. Not only did nature sustain existing life-forms, it repeatedly demonstrated the ability over geologic and evolutionary time to generate more life in the face of adversity. Producing numerous novel species, even after massive disturbances and die-offs, nature tended to add complexity to its products as well. Something in the system seemed to be endlessly generative. Since we clearly valued the species produced, Rolston thought we must also value the system responsible for producing them. He labeled this remarkable creative phenomenon earth's "systemic value."

At the end of his analysis, Rolston believed he had described a planet replete with natural value at every level. Intrinsic and systemic values were the necessary substrate for all instrumental valuing of nature. The old assumption that all value required a human valuer appeared now implausibly parochial, a misguided assumption made by a species arriving, as the paleontologists noted, rather late on the planetary scene. Rolston was careful not to deny the significance of humans and the things they held in high regard. Humans were a particularly notable product of the system. What they valued mattered. At the end of the day, however, *Homo sapiens* represented just one very impressive, recent evolutionary story among many other remarkable ones. Every creature was a living example of the relentless accumulation of natural value over geologic time.

Now seven or eight years into his job in Fort Collins, Rolston was beginning to feel almost as much at home among the Colorado flora and fauna as he had in Virginia. For the first few years in Fort Collins, he had sent out applications each fall for jobs in the Southeast and East, hoping that he and Jane could move back to somewhere more familiar, perhaps somewhere closer to those "rarified scholastic atmospheres" he had earlier coveted. As time wore on, Rolston found that the ecological drama of the western landscapes was making its way deeper into his soul. The less populated Colorado landscapes seemed to give clearer expression to the natural forces that lay at the center of his thinking. When back East, he noticed he increasingly needed his guidebooks to remind him of the names of the plants and birds. Though he still felt a strong emotional pull from his native Virginia, he knew he was starting to belong in Colorado.

Rolston and Jane were by this time raising a family in Fort Collins.

Their two children, Shonny and Giles, adopted in 1969 and 1972, were starting to come on hikes and camping trips with their parents. Rolston bought a Steury Camper to tow behind the car for weekend excursions. As his father and grandfather had done before him, Rolston tried to encourage in his children an interest in the natural world. Passing wisdom and know-how down through the generations was for Rolston not just part of a fancy philosophical argument about DNA and moral value, it also the reality of his life as a parent. Giles was increasingly his sole companion on many of the hikes, showing a toughness that made Rolston proud. By now it was obvious their children would grow up as westerners, something even Jane was beginning to accept.

The experiences Rolston valued most during his outdoor explorations always involved the poignancy of life's struggle for survival. He described in his trail log the day he struggled on snowshoes against bitter wind and cold through a stand of willows at Guanella Pass, high in the central Colorado Rockies. Close to the top of the pass he came across a ptarmigan camouflaged white against the snow, perched in the lee of a rock. Until it moved, all that was visible was a small black eye and bill as the bird crouched low against the blowing snow. He found himself moved by this struggle for survival in the face of the elements: it cut to the heart of how nature created value. Life, he noted, is "pressed by the storms, but it is pressed on by the storms . . . environmental necessity is the mother of invention." The creativity inherent in the system constantly produced mechanisms in response to threats, in the process creating life-forms with both function and beauty. Set against the relatively temperate Shenandoah, the bitter winds and snows of the Colorado high country made mere survival seem all the more remarkable.

The decision to leave the pastor's life nearly a decade previously had been motivated in part by the need to work through the intellectual puzzles, a process which by now he felt was yielding some good results. But his departure from Virginia had also been motivated by the intention to create real change in the world. As his professional career started to take off, Rolston felt the urgency to become more active in protecting the parts of nature being lost. He was still haunted by memories of the destruction of his native Virginia landscape. Similar destruction was now becoming evident in parts of the Rocky Mountain west. Colorado was filling up with recreationalists and retirees. Every month, it seemed, developers were carving new subdivisions out of horse pastures on the edge of town. Rolston felt the need to do more than just play with words for the edification of sophisticated intellectual audiences. Philosophy needed also to be practical. As he gained a reputation as an innovative thinker on environmental issues, he looked for places to employ some of his arguments on the ground. He didn't have to look far.

Rising to the west of Fort Collins was a distinctive notched foothill known as Horsetooth Mountain. Three main rock outcrops formed the peak, separated by two deep clefts through which the light shone. The gap-toothed formation made the location of Fort Collins easily recognizable from the plains to the east and from the air. To residents of the city looking up from below, Horsetooth was a herald of the much taller and more imposing landscapes that lay to its west.

In 1980 Johnny and Virginia Soderberg, the ranchers who owned Horsetooth Mountain, tried to find a buyer for their property. The reservoir beneath Horsetooth made this ranchland into prime real estate for developers of high-end housing. The owners were willing to sell to the county rather than to a developer, but the county needed to raise the money for the purchase. Someone proposed a penny sales tax to fund the purchase of Horsetooth. It would first have to be passed by citizens usually reluctant to burden themselves with additional taxes.

Rolston knew there were natural values of all kinds at stake in and around Horsetooth. It was a close-in recreational site visible from anywhere in town. The land was ecologically interesting, occupying a transitional zone between the montane ecosystem to the west and the grasslands to the east. It offered a small taste of the geologic drama responsible for the Rocky Mountain high country. Trails wound their way to the summit through ponderosa pine, mountain mahogany, and rocky mountain juniper shrubs. Up to forty species of birds called Horsetooth home in summer and large mammals such as black bear and mountain lion padded regularly through its rock-strewn forest. Only four miles from town, Horsetooth was a place where Fort Collins residents could come and experience wild nature and the forces that had shaped it.

Though temperamentally more inclined to steer clear of politics, Rolston used his pen to quietly persuade his fellow citizens of the importance of a yes vote on the sales tax referendum. In an opinion piece for the *Fort Collins Coloradoan*, he wrote that Horsetooth "forms our western skyline, framing a skein of geese or a wintry sunset, to add a touch of the wild to our routine town." The mountain was a living symbol of Colorado's geologic history. Though the rocks themselves might survive development, the essence of the historical landscape would not. Subdividing the land into ranchettes for the benefit of a wealthy few would, claimed Rolston, cause Horsetooth to be largely disfigured, with the wild and unspoiled effect being lost forever.

Rolston had started to develop certain principles for applying his environmental theory to policy. He used these principles now to persuade people to vote yes on the sales tax. Good environmental policy should preserve those parts of nature that are important cultural symbols. It should recognize the increased value of wild areas easily accessible from town. When difficult choices had to be made, policy makers should favor nonconsumptive uses such as hiking over destructive and irreversible uses such as housing construction. All of the principles pointed toward spending the money to save Horsetooth from development.

In April 1981 the sales tax passed by a small margin. The county became owners of the new park and residents of Fort Collins started to enjoy a network of trails close to town. Even residents living in other parts of the county evidently answered Rolston's plea to "be good neighbors." Having chosen to tax themselves for a long-term environmental good, citizens ensured they would forever gaze up at an important symbolic and recreational site unmarred by roads and ranchettes. On this occasion, environmental ethics had shown its power over economics.

Just three years after the battle for the future of Horsetooth Mountain, the Poudre River to the north and west of Fort Collins was threatened with a dam. Boosters claimed that damming the Poudre would provide water during the dry summer months for ranchers at the bottom of the canyon. A hydroelectric project would also provide year-round power to the local grid. Environmentalists pushed in the other direction. A wild Poudre was worth preserving. Too many natural values were at stake. At the same time as campaigning against the dam, the environmentalists started lobbying the U.S. Congress for the river's wild and scenic designation.

Rolston saw this as a fight worth joining and again used all the arguments in his environmental ethics arsenal to persuade his fellow citizens to protect the Poudre. He made the case in local papers and newsletters that a dam was not just unnecessary but also morally wrong. Damming the wild Poudre would be an affront to the natural values present in the canyon. Rolston urged citizens to contact their senators and representatives.

Rolston acknowledged that those who wished to preserve the canyon found it hard to articulate exactly what it was they wanted to protect. The preservationists seemed to have only "soft" values on their side as opposed to the "hard" demands for water and power highlighted by development advocates. These soft values, Rolston argued, were important ones. He claimed that a preserved Poudre would provide not just recreational but also geologic perspective to an increasingly urban people. The canyon was, in Rolston's words, "the cradle of thoughts and aspirations that individuals and society cannot afford to do without." The age-old gorge and the river-cut strata provided "insight into who and where we are in the earthen world, a sense of human transience and of the perennial, encompassing natural certainties." He argued that the preservation of such symbols of time and eternity would make Coloradans better global citizens. Wild and free rivers were an irreplaceable gift of nature to society. There existed few better indicators of the tensions between movement and stability, permanence and flux that had shaped the history of life on earth. People needed exposure, Rolston urged, to those systemic forces.

In 1986 a bill introduced by Colorado Senator William Armstrong proposing wild and scenic river status for thirty miles of the Cache la Poudre—and recreational status for forty-six more—passed the U.S. House and Senate before being signed into law by President Reagan. The Poudre became the first river in Colorado to gain federal Wild and Scenic designation. Rolston's environmental ethic seemed to have resonated with a popular sentiment about the Poudre. Though he did not delude himself into thinking his arguments were the decisive factor, his position seemed to strike at some hidden chord with his neighbors. His approach to environmental value was not just a clever piece of academic fancy footwork. It also had traction on the ground.

Gaining confidence now that his position had proved popular with the public, Rolston continued to refine its subtle structure. Academic philosophy had for centuries worked by laying down challenges to the works already in print. Since Rolston's argument about instrumental, intrinsic, and systemic values had now appeared in several journals, he found himself under increasing pressure to elaborate and defend his views. While many progressive thinkers were open to the idea that living beings like whales and bears might possess intrinsic value, the claim that whole ecosystems possessed such value remained much more controversial. Ecosystems were chaotic, disparate entities, lacking the kind of unity that seemed necessary to make them worthy objects of respect. Since ecosystems were made mostly of components not typically thought to be alive, such as mountains, water, and soils, the old problem of finding moral value in dirt resurfaced.

To answer this puzzle, Rolston brought the focus off the rocks and the ecosystems and onto the creative forces that lay beneath. He returned to the idea that earth contained stories. The idea of story had been central to his view that DNA recorded achievements in living organisms over evolutionary time. Rolston knew there was also storied history at work on Horsetooth Mountain and in the Poudre River. He now worked to articulate the underlying forces propelling geologic and evolutionary history forward, noting how certain countervailing tendencies seemed to be the core of nature's productivity.

He drew attention to the ever-present tension between stability and spontaneity in nature. There was a continual swing in biotic systems between the regular and the unpredictable. Tides were regular, rains tended to fall more in certain months of the year, and temperatures historically ranged between established limits. In contrast to this relative stability, earthquakes and twisters occurred suddenly and lightning struck unpredictably, sometimes occurring in just the right place to ignite a forest fire that transformed a landscape for a century. In the Mount Zirkel Wilderness, Rolston had hiked for a full day through the remnants of a forest blown down by a once-in-five-hundred-years wind event. A few years later, his basement office at CSU was filled to the roof with water after ten inches of rain caused a tiny campus creek to burst its banks. Stability alternated with spontaneity to create openings and opportunities for nature's productive powers to work their effects. As a result of the flood, his office was ruined, but the foothills flowered the next spring as never before.

Next Rolston articulated the "diversity-unity" value of systemic nature. He found this value in the contrast between the relatively few ultimate constituents of nature and the myriad of different life-forms those constituents created. DNA and RNA, for example, were just two types of molecule but they were found in every form of life. These two types gave shape to all of earth's biota, from staphylococcus bacteria to sperm whales. Similarly, just three types of particles—protons, neutrons, and electrons—made up all the physical matter in the universe, those basic elements coming together in an infinite number of different ways. The relationship between the one and the many was one of earth's perennial mysteries. Relative singularity was somehow able to spawn incredible diversity.

Throughout biological and geologic nature, Rolston saw instances of opposite forces working productively against each other. Stability was in tension with spontaneity. Diversity was in tension with unity. Nature sometimes followed law-governed necessity and sometimes followed chance. Oscillations between plenty and scarcity drove animals to adapt creatively. The system was driven by an endlessly creative tension, with all of these opposites propelling earth's history dramatically forward. The result was a diverse, complex, and fertile earth, apparently unique in the known universe.

Rolston knew his arguments in support of the systemic value of nature were not expressed in customary philosophical form. But there was something about earth's history Rolston found consistently eluding logic. Rather than arguing like an engineer, fitting together precisely the pieces of a complex philosophical puzzle, Rolston reached toward an articulation of some deep underlying force responsible for earth's remarkable story. He explained this to his fellow philosophers as best he could.

> Bioscience can present little theoretical argument explaining this history—little logic by which there came to be primeval Earth, Precambrian protozoans, Cambrian trilobites, Ecocene mammals, Pliocene primates, and Pleistocene *Homo sapiens*. . . . Likewise, passing from science to ethics, philosophy can present no argument why these stories ought to have taken place. . . . But we can begin to sketch nesting sets of marvelous tales . . . and hope you can accept them for that.

With ecological and evolutionary science making people increasingly aware of the "marvelous tales" to be told about the natural world, Rolston's unusual style of philosophy was gaining more adherents with every article he published. His fears about whether his work would ever be appreciated when he made the change from pastor to philosopher began to fade as the long evenings spent in his philosophy office started to pay off.

Initially, it had been the natural resource specialists who most quickly embraced his ideas about the systemic value of nature. They were thrilled to encounter a philosopher who could articulate the moral value of the natural web they studied and they were perhaps less aware of the philosophical unorthodoxy of the view. Now growing numbers of serious philosophers were also being drawn in by Rolston's lyrical arguments in support of nature's intrinsic value. Ethicists across the region and the country were increasingly likely to be discussing Rolston's new book *Philosophy Gone Wild* and assigning it in their classes. Even his previously skeptical colleagues in the Colorado State philosophy department could not help feel some admiration for the stir their disciplined and hardworking colleague was beginning to make. Maybe Rolston was sermonizing, but clearly people wanted to hear the message.

Though the pressure was diminishing, in order to appease the philosophical mainstream Rolston remained meticulous about ensuring that the words he used had a strong secular appearance. By all accounts, he was extraordinarily successful with this tactic. But when Rolston started talking about deep creative forces inherent in the system, it did not take a particularly perceptive observer to detect the waves of theological resonance still reverberating deep within his thinking.

9

Ecology and Aesthetics in Yellowstone

> On the stage of natural history, the human phenomenon is ephemeral. . . . Wildlands provide the profoundest museum of all, a relic of the way the world was 99.9 percent of past time. . . . An immense stream of life has flowed over this continent Americans inhabit, over this Earth . . . people of every nation need nature as a museum of what the world was like for almost forever, before we so recently came. [*1988*]

THE PHILOSOPHERS ROLSTON TOOK hiking into the Colorado high country could not help being impressed by his obvious expertise in the natural sciences. Rolston hiked with binoculars and a hand lens strung around his neck, a notebook, and a makeshift plant press in his pack. Reference works came with him in the car and many more were stacked on the bookshelves back home. Few university professors could claim such genuine interdisciplinary competence. To maintain this edge, Rolston continued taking classes from his colleagues in the science departments at Colorado State, sitting in with students less than half his age.

One summer away from teaching, he spent three weeks in the White River National Forest with William A. Weber, curator of the herbarium at the University of Colorado, learning the flora of the western slope. In 1982 he published the discovery of a new moss previously unknown in Colorado, *Bryum knowltonii*, in the journal *Bryologist*. Another of his moss discoveries, *Cynodontium gracilescens*, was one of few records in North America. On a prompt from a local expert, he backpacked three days into remote kettle ponds in the Mount Zirkel Wilderness to find *Drosera rotundifolia*, a rare insectivorous species left over from the Pleistocene period. Over the years, Rolston had clearly become an accomplished Rocky Mountain naturalist.

Though the central pivots of his environmental ethic—notions such as systemic and intrinsic value and themes such as story and creative tension—required no particular geography for their application, they did require a deep appreciation for evolutionary and ecological processes. Often these processes were more evident on the less-impacted western landscapes. The examples and illustrations Rolston used in his arguments—appreciating the pasqueflower, setting aside wilderness, letting rivers flow free, restoring wolves—were drawn largely from the biology and ecology of the western mountains. Hence it was little surprise that, before long, he was deeply involved in a dispute over one of the most famed of those western landscapes, Yellowstone National Park.

In two years in the late 1980s, a pair of events generated more serious questions about the preservation of Yellowstone National Park than anything else during the previous century of the park's existence. Both events focused attention on the question of how much the Yellowstone ecosystem should be allowed to run wild. In 1986, philosopher and ecologist Alston Chase published *Playing God in Yellowstone*, a book containing a sharp indictment of the National Park Service policy in Yellowstone's ancient caldera. Chase's book documented the history of the management of Yellowstone since it was first set aside as a park in 1872. He cataloged a number of what he saw as Park Service failures. These included the mismanagement of the elk and buffalo herds, the eradication of wolves and other large predators in the 1920s, the failure to protect grizzly bears from the closure of Park Service dumps in the 1970s, and the recent building of Grant Village on Yellowstone Lake. According to Chase, the politicians and the bureaucrats in charge of this national treasure had done immeasurable harm to the park's ecosystem.

While a number of the failures Chase pinpointed were cases in which the active management of park lands appeared to have gone wrong, the real target of his book was the guiding park ideology of "natural regulation." This Park Service policy reflected the underlying idea that nature knows best. Human interference with ecological processes was undesirable. Yellowstone could come to no harm, advocates of natural regulation assumed, provided humans simply stood back and let the park's ecology manage itself.

The policy of natural regulation had been instigated after the publication of a 1963 Park Service document known as the Leopold report. The report had been written by Starker Leopold, whose father, land ethicist and game management specialist Aldo Leopold, was author of the landmark book *A Sand County Almanac.* Commissioned primarily to address a problem with overgrazing, the Leopold report recommended initially reducing the elk herd in order to recreate the ecosystem present in Yellowstone at the time of European colonization. After this short period of active management, nature should then be left alone to regulate itself. Starker Leopold envisioned a park eventually operating in something close to its primeval condition. The Park Service accepted the recommendations of the Leopold report and natural regulation had been the guiding management policy ever since. The contentious ecological claim behind the policy was that the Yellowstone area was big enough and its systems intact enough to regulate their own health. This, according to Chase, was one of the Park Service's biggest mistakes.

The ecological assumptions, Chase argued in his book, were only part of the management disaster. Natural regulation had become not just a scientific position but also an ethical one. "Let nature run its course" had become an unassailable mantra treated by the Park Service as if it represented the moral high ground. The most morally pure management plan the Park Service superintendents could adopt was one of no management at all. This had become virtually a first commandment of Yellowstone.

Chase argued throughout his book that this was both bad science and bad ethics. The park was not big enough and the ecological systems not intact enough for such a policy to work. Self-regulating nature was nowhere to be found on the Yellowstone landscape. Beginning with the active burning and hunting in the park by Bannock and Shoshone peoples, Yellowstone had long been subject to human influence. There had been no balance of nature without humans for tens of thousands of years. People were today as integral to Yellowstone ecology as elk and grizzly bears. The ecosystem was neither self-contained nor self-regulating. From his own observations as a historian, a scientist, and a Park Service veteran, Chase was convinced that the laissez-faire attitude of natural regulation was leading to the slow death of the park's flora and fauna. He claimed to have written the book out of a sense of despair, stating that he could not stand to see Yellowstone killed by a mistaken environmental ideal. Chase's book immediately put the Park Service on the defensive.

The second big event in Yellowstone in the late 1980s attracted far more national attention than Chase's philosophical pleading. This time nature itself directly posed the questions. On June 22, 1988, a lightning strike ignited a small fire in a stand of lodgepole pines in the northwestern corner of the park. According to the Park Service's natural burn policy, fires started by lightning were left alone so they might play their historic role in ecological succession. This was natural regulation at work. Rains that normally fell during June were expected to keep the fires in check.

The summer of 1988 was shaping up to be one of the driest and windiest on record. Instead of burning for a few hundred acres then extinguishing itself as expected, the fire, fanned by the powerful winds, ignited more of Yellowstone's backcountry every day. Further lightning strikes added to the flames. By late July, close to 50,000 acres of Yellowstone were burning with no relief from the hot and dry weather in sight. A chorus of voices against the Park Service natural burn policy began to rise. This great cathedral of the West was being destroyed by wildfire.

Through July things only got worse. In response to increasing public outcry and pressure from the U.S. Congress, Yellowstone reversed its natural burn policy on July 22, and started actively fighting the fires ravaging the park. But by this time, the fire line was wide enough and the burn hot enough that fire suppression had become an almost impossible task. The Yellowstone inferno became the largest fire complex in the recorded history of the region. A total of 9,500 firefighters with over one hundred engines and a similar number of aerial tankers were brought in to combat the conflagration. Eight hundred miles of fire line were constructed throughout the Yellowstone ecosystem, over a million gallons of fire retardant were dropped, and $120 million of taxpayer monies were spent before the early autumn snows finally doused the flames. By the end of the fire season, over 1.2 million acres of the Yellowstone ecosystem had burned, 70 percent of those lying inside the park. Sixty-seven structures had been destroyed. A third of America's most famous national park lay smoldering, reduced to ash and embers.

Large portions of the American public, spurred on by a national press gorging on the controversy, were outraged. The Yellowstone fires received more media coverage than any other event in national park history. How could a government agency charged in the Organic Act of 1916 with protecting and preserving the park in order to leave it "unimpaired for the enjoyment of future generations" sit by for a month and watch with apparent indifference while a national treasure was ravaged? The guiding policy of natural regulation governing the management of Yellowstone suddenly looked like a cover for the inaction of fools. The *Richmond News Leader* advised its readers, "If you want to see the world's largest charcoal grill, just visit Yellowstone." The *Billings Gazette* claimed the park had ridden "a dead policy into hell." Many observers were outraged at how the park appeared to have been permanently disfigured as a result of a misguided government policy. The idea of natural regulation was taking a battering and Chase was having a field day as his book flew off the shelves in bookstores across the country.

Nineteen eighty-eight was not just the year Yellowstone burned, it was also the year Rolston published a systematic work outlining his ethical arguments for protecting the natural world. The philosophy articulated in *Environmental Ethics: Duties to and Values in the Natural World* offered Rolston's most detailed arguments so far about the intrinsic value of natural systems. Its key message stood almost diametrically opposed to that of Chase's accusatory tome. Natural values were inherent in the Yellowstone ecosystem. Human intervention by heavy-handed, if well-intended, managers, risked reducing the historical integrity of the landscape. The Park Service managers who supported natural regulation were looking for someone who could take on the ideas behind Chase's arguments. Rolston suddenly found himself recruited into an uncomfortably public confrontation.

Neither Rolston nor Chase had written their books intentionally to target the other's position. Nevertheless, many of Rolston's arguments could be applied to that use. In fact, several of the examples Chase used in *Playing God* to castigate the Park Service were used in the opposite way by Rolston to illustrate how an ethic embracing the intrinsic value of natural systems might be put into practice. For example, in the case of the Yellowstone bighorn sheep herd stricken with treatable chlamydia, Chase had excoriated the Park Service for being unwilling to treat the sheep to cure their blindness. He was appalled when 60 percent of the park's bighorn herd lost their sight and died in just one winter, noting the bitter irony that some of these sheep, after wandering on the roads and being struck by cars, were mercy-killed by the same officials who had refused to treat their eye disease. Rolston, on the other hand, cited this hands-off policy as a good example of letting evolutionary forces do their work. In the long run, Rolston argued, natural regulation would allow the herd to become resistant to the disease through selective processes.

Chase also vilified the Park Service employee who watched impassively while snowmobilers tried to rescue a bison that had fallen through winter ice. The employee cautioned the would-be Samaritans to let nature take its course. This time the park officials would not even mercy-kill the animal. Chase was appalled at the cruelty.

Rolston cited the same incident as an appropriate natural regulation strategy for the park. Cautious bison would survive thin ice, while the foolhardy ones would not pass on their genes in the herd. Leaving the bison to drown was the morally correct choice. According to Rolston, "Let nature take its course" was in Yellowstone a more appropriate ethic than "Be kind to animals." The natural order sometimes needed protecting more than the well-being of any individual bison.

In an article published in the journal *Philosophy and Biology*, Rolston entered the debate over Yellowstone's management more directly. He accused Chase of being "relentlessly one-sided and mean-spirited." Rolston argued that while it would be an illusion to think of the park as entirely free of human influence, it was appropriate to have a management policy that minimized the scale of human impacts. Letting nature take its course, he suggested, "is the appropriate way to show care for the great bear, the wapiti, the aspen, *Ursus arctos, Cervus canadensis, Populus tremuloides*, for the ecosystem, for the land, for this wild place." Though Rolston's article was published in a professional philosophy journal, it was quickly picked up and passed around the Park Service. The gist of it was reprinted in the regional paper *High Country News*. Natural regulation, the Park Service could now say, was an appropriate strategy, ethically sanctioned by a philosopher of national repute.

Invited to give a keynote address at the Greater Yellowstone Coalition meeting in May of 1989, Rolston pulled off the gloves. He pointed out that while it was possible to get lost geographically in Yellowstone, Alston Chase had also managed to get himself lost philosophically. The type of intense scientific management Chase supported for Yellowstone would result in a park more like a zoo or Disneyland than an example of raw nature. Visitors would admire the skills of the park managers rather than the values present in the natural system. Using arguments familiar from the Horsetooth Mountain and Poudre River battles, Rolston characterized Yellowstone as an invaluable symbol and a near-pristine archetype, one that was becoming increasingly scarce throughout the developed world. Rolston made the case that to interfere with such an archetype was not only biologically unnecessary but also morally wrong.

In response to Chase's assertion that Yellowstone had always been altered by humans, Rolston acknowledged how in times past humans might have been heavy-handed in their management of the park. This did not, however, mean Yellowstone was no longer predominantly wild. The emphasis Rolston placed on the perpetually present creativity of spontaneous nature meant that, once left alone, "natural forces will flush out many human effects." At the end of the day, nature comes back. In Yellowstone, Rolston insisted that human impacts were minimal enough for nature to still be acting independently there.

> "Wild Nature" successfully denotes, outside of culture, an evolutionary and ecological natural history, which remains present on the Yellowstone landscape, though jeopardized by numerous human influences. . . . Natural processes can be preserved today because of, rather than in spite of management policies.

The much-maligned management policies to which he was referring were those of natural regulation.

The lengthening perspective in the decades since the fiery summer of 1988 appeared to vindicate Rolston's position. Park visitors found a vigorous regrowth of vegetation taking place in the "world's largest charcoal grill." The ecosystem as a whole, far from being destroyed by the fire, appeared to prosper as a result of it. Wildlife mortality was remarkably slight, with fewer than four hundred large mammals perishing from the smoke and flames. Most of the animals killed were elk, with only nine bison and one grizzly bear perishing in the fire itself. Large numbers of additional deer and elk starved the following winter due to a decrease in available forage, but the forbs and grasses bounced back with vigor the following spring and the deer and elk population quickly recovered. Many seeds lying dormant underground were undamaged, since the scorching of the soil penetrated less than two centimeters in most places. Even in the most severely burned portions of the park, most of the vegetative root systems were left untouched. The flood of nutrients in the ash and the increased solar radiation reaching the forest floor ensured that the following spring produced a profusion of wildflowers. The differing intensities of the fire left behind a stimulating ecological patchwork.

In comparison to the explosions of the ancient caldera that had occurred twice in recent geologic history, the fires of 1988 were in fact a relatively small disturbance. Nature's creative forces had thrived under these types of disruptions for hundreds of thousands of years. Even a "catastrophic" fire, happening on average every three hundred years, turned out to be an essential part of the native processes. Rolston's ethical perspective enabled him to quickly appreciate how the disorder caused by a fire, a flood, or a sudden windstorm had to be set inside the context of the productivity inherent within a large ecological systems.

Though the debate about Yellowstone policy continued, Rolston found his own reputation enhanced by the events of 1988. His philosophical arguments had given credibility to the Park Service policy at a time when the policy sorely needed it. The park superintendent, Bob Barbee, who had come under enormous pressure after the fires, personally thanked Rolston for offering an ethic that so eloquently supported the park's policy of natural regulation. In turn, the Yellowstone fires offered a test case for Rolston's philosophy. When the smoke finally cleared, his philosophy seemed to have passed the test.

The large-scale destruction of any natural environment by fire, hurricane, or flood typically causes at least two types of negative reaction. One is an ecologically motivated reaction such as Chase's, the worry about the system's ever regaining its complexity and vigor. The other is a reaction based in aesthetics. One of the primary reasons for the uproar when Yellowstone burned was the feeling that a nationally known landscape of sublime beauty was being replaced by an ugly scene of charred pines and ash.

Smokey Bear had for decades been urging Americans to keep their forests green. At the end of the 1988 fire season, very little of Yellowstone was even remotely green. The photographs taken by the few tourists who dared enter the park that fall were not taken to impress the folks back home with the unrivaled beauty of the ancient caldera. They were taken to illustrate what an unqualified aesthetic disaster had befallen one of America's crown jewels. A blackened Yellowstone Park was ugly. Ugly was not something you expected to find in a national park.

Forty years previously, when Oxford University Press posthumously published the book many thought anticipated modern environmental ethics, Aldo Leopold had given a hint of why beauty mattered. In one key sentence toward the end of his *Sand County Almanac*, Leopold tried to capture the motif at the heart of his land ethic, the essence of what made nonhuman nature morally considerable: "A thing is right when it tends to preserve the integrity, stability, and beauty of the biotic community. It is wrong when it tends otherwise."

In phrasing things this way, Leopold stipulated a powerful criterion by which to characterize natural value. In the process, he also unwittingly indicated just how close to natural value stood aesthetic value. Natural environments possessed not only ecological integrity and stability, they also possessed beauty. Beauty, according to Leopold, was morally significant too.

By this time Rolston felt a strong affinity with Leopold's thinking. He had recently spent an afternoon at the great conservationist's shack in Wisconsin's Sand Counties. Afterward he wrote fondly in his trail log about the occasion.

> *Tuesday, November 15, 1988.* [Nina Leopold Bradley] reminisced about her father, Aldo, while about a dozen of us were inside—hard rain outside and some lightning. I sat in a lawn chair that Leopold had made. A very plain shack but she has kept it in vintage condition with the implements, furnishings, etc. that were there when Leopold was there.

Leopold's championing of nature's integrity, stability, and beauty seemed to Rolston to hit the mark.

One would be hard pressed to find any environmentalist who doubted that beauty was an important part of the reason for protecting nature. Yellowstone and Yosemite National Parks had gained protection at the end of the nineteenth century as much for the visual feast they offered the tourist as for the exceptional ecological and geologic values they contained. Similarly, it was the majestic mountain scenery that brought visitors to Rolston's home state of Colorado to take photos and relax in front of picture-postcard vistas.

While it captured something important about why we should care for nature, Leopold's statement about "integrity, stability, and beauty" also introduced a philosophical problem. He had used "integrity" and "stability" as value words, but at least they referred to a pair of more or less objective characteristics of an environment. A scientist had criteria for measuring a system's integrity or stability. Beauty, on the other hand, was a much more subjective category. Whether or not something counted as beautiful was a judgment made by a human observer. Opinions differed. Beauty, as everybody well knew, was in the eye of the beholder.

Leopold's statement left environmentalists with a catalog of problems. If beauty was part of what gets environments protected, who were the appropriate judges of what counted as a beautiful environment? Perhaps a western landscape artist would celebrate untouched mountain scenery, while a city dweller might find the same view more pleasing with a restaurant and some shops nestled at the base of the peak. Leopold's use of the term beauty seemed to imply that ugly environments were not worth protecting. At the start of the environmental movement, some natural systems, such as deserts and prairies, suffered severe image problems. There were very few advocates lobbying for their protection on account of their beauty. Using Leopold's criteria, their perceived ugliness was a liability. A third problem arose when one looked ahead. What would happen to the environmental movement when all the pretty landscapes had been protected? Should society then draw up a secondary list of "somewhat beautiful" landscapes?

Rolston knew that aesthetics had to remain in the taxonomy of environmental values. He also well understood the potential pitfalls for environmentalists. The Rocky Mountain peaks, Rolston realized, were thought by most people to be spectacular and worthy of protection on aesthetic grounds. Yet their bare summits, sheer cliffs, and talus slopes lacked the more biologically complex ecology of the less eye-catching valley bottoms. From the perspective of the integrity and stability of the system, the valley bottoms might be more important to protect than the spectacular high country. From an ecologist's perspective, beauty could be a fickle guide.

As a man who had spent countless hours of his youth wading through the Florida and Alabama swampland, Rolston was sensitive to how Leopold's aesthetic criterion created particular difficulties for landscapes such as swamps and marshes. The word "swamp" was almost synonymous with the idea of ugliness. Swamps were perceived as dark, dank, disease-ridden, overgrown, and gloomy. One East Coast area between North Carolina and Virginia had been christened "The Dismal Swamp" as far back as 1669. Rolston owed it to his southern roots to show why it was wrong to discount the moral significance of swamps. The challenge, as he saw it, was not to deny that aesthetics provided good grounds for protecting nature. Clearly it did. It was to figure out how even those parts of nature traditionally deemed ugly had aesthetically positive dimensions to them.

Rolston was struck by the remark of a historian writing about England's eastern marshlands: "Any fool can appreciate mountain scenery. It takes a man of discernment to appreciate the Fens." Rolston had spent more than enough time down on his knees with a hand lens in soggy woodlands to know what this Englishman meant. Ecologists knew that wetlands were some of the most biologically diverse and productive ecosystems of all. Finding beauty in swamps required a blend of more biological science and a more mature aesthetic sensibility. Rather than accept the ugliness of swamps, Rolston recommended a closer look at their ecology. In essence, he suggested that the philosophers of aesthetics needed a science lesson.

If one looked at swamps with an eye on the catalog of natural values Rolston had articulated, a remarkable thing happened. These dismal landscapes start to overflow with a profusion of natural values. Swamps and wetlands were biodiversity hotspots, unsurpassed ecological sanctuaries for flora and fauna. The water stored and filtered by a wetland had immense instrumental value as the lifeblood of the terrestrial biota. Startling and rare plants like the carnivorous Venus flytrap and the pitcher plant made their homes nowhere else but in select swamps.

Swamps and wetlands were particularly dynamic landscapes, some of them disappearing entirely by midsummer only to be replenished in a different shape and with a new roster of inhabitants by the fall rains. The mixture of order and disorder in a wetland could be fiendishly productive. Marshes were unified and diverse, stable and spontaneous, ephemeral and long-lasting. The complexity of a swamp was evident in the range of organisms coexisting there. A raft of nearly identical pond lilies could be suddenly pushed aside by the sweep of an alligator's tail. Each inhabitant of the swamp was continually challenged to find a way to live successfully in its changing niche. The struggle to survive so evident throughout this watery world was a source of immense creativity. Swamps contained some of the best illustrations of nature's systemic values at work.

Rolston recognized how an appreciation of swamps required a particular openness to the role of death in the system. The bottom of a wetland could be filled with a century's worth of decaying matter. A handful of this sludge revealed a myriad of decomposing plants and animals dripping through one's fingers. The decaying matter supplied the nutrients that made the wetland so biologically rich. Swamps therefore brought into especially sharp relief the role of death in propelling life forward. The immense sweep of natural history buried within a swamp became the underlying force that drove the surface life forward. Rolston worked hard to persuade people of these hidden powers:

> After one has become ecologically sensitive, the system takes on the qualities of a kaleidoscope, in which the accidental tumbling of bits and pieces, each with its own flash and color, constitutes a set of patterns of interdependent parts coacting, patterns repeated over time and topography, endlessly variable and yet regular, buzzing with life. . . . Survival—making it through, living on and on—is the last word, life's deepest beauty.

When one recognized the historical processes responsible for this snapshot, it was hard not to see beauty where one could not see it before. The key to finding the swamp aesthetically stimulating, Rolston stressed, was to move beyond the traditional aesthetic notions of the scenic and the pretty toward something more ecologically informed. Aesthetic sensibility was heightened by a grasp of the science.

The same method used to appreciate swamps worked with other natural landscapes not previously considered beautiful. These included landscapes like the prairie, the tundra, the desert, and even the charred forest. In each case, an individual's aesthetic sensibility required education and adjustment to appreciate the alien scene. Sensory prejudices had to be unlearned and new sensibilities mastered to appropriately experience the new environment. As Rolston and his wife had learned, those raised in the warm and humid ecologies of the Southeast had to shift their habits of observation when traveling across the hundredth meridian. The panoramas available in Rolston's adopted Rocky Mountain home were dramatically different from those in the landscapes of his Shenandoah Valley youth. The tan-colored earth of the Colorado Front Range sprouting bunch grasses and silver-gray sage offered strikingly different vistas from the multilayered florescence of Virginia. Aesthetic expectations needed to be adjusted. Similarly, most residents of the Rocky Mountains traveling East needed to be educated into the aesthetics of southern swamps.

One hundred years earlier, having hiked across much of the Sierra Nevada, southeast Alaska, the southern Appalachians, and the Mississippi Valley, environmentalist John Muir made a startling claim. "None of nature's landscapes are ugly," he claimed, "so long as they are wild." Muir's view later became known as "positive aesthetics," the idea that all natural landscapes have positive aesthetic value. Rolston's approach to environmental aesthetics was to argue that this positive evaluation hinged not so much on a preprogrammed emotional response to nature as on a more informed understanding of the landscape. All wild landscapes, if understood in their proper scientific context, were beautiful landscapes.

An explosion of invitations to speak across the country gave Rolston the means to cultivate his own aesthetic sensibilities. One of the conditions of his speaking engagements was that he would come a day or two early or stay on over the weekend, and that his hosts would find a knowledgeable biologist on campus to take him out onto the landscape. The philosopher found himself with increasing opportunities to experience the aesthetics of a diverse range of landscapes.

Rolston's position in aesthetics, like his position in ethics, was heavily informed by the natural sciences. His expertise there was continuing to feed his philosophy. The connection between science and ethics that his colleagues had warned him against when he first arrived in Colorado was turning out to work to his advantage. The more science he put into his arguments, the more compelling his audiences found them, and the more stimulating they, in turn, found their landscapes—not only in the Rocky Mountains, but also in environments traditionally not thought of as beautiful.

By the early 1990s, environmental philosophy was firmly established in the academic landscape of the United States. Over three hundred schools and colleges offered at least one undergraduate course in the subject. Graduate programs in environmental philosophy had started in Colorado, Ohio, Texas, and Montana. Rolston found himself at the epicenter of a rapidly growing field. His books were selling well. His papers had been reprinted over a hundred times and were being read around the country. The Colorado Division of Wildlife, the Environmental Protection Agency in Washington DC, and the U.S. Forest Service had all sought out his expertise.

Rolston knew from his training that good philosophical arguments, like good scientific generalizations, should transcend geography. The arguments for intrinsic value he had been making for more than a decade were designed to work not only outside the Rocky Mountain west, but also outside the United States. The reasons it was morally wrong to destroy a river delta on the U.S. gulf coast were essentially the same reasons it was wrong to destroy one in Bangladesh. The connection between the system's natural value and biological generativity meant his arguments should work wherever biology worked.

But when Rolston's environmental ethic started to receive international scrutiny and he began to get invitations to speak overseas, a strange thing happened. At the same time as he was being feted at home, somewhat to his surprise and much to his dismay, Rolston found himself facing an increasingly hostile reception overseas. His Rocky Mountain ethic was apparently not as ready to go global as he had hoped.

PART IV

Global Environmental Ethics

10

Wilderness and the Developing World

> Wild nature is a place of encounter where humans go, not to act on it, not to labor over it, but to contemplate it. . . . Wilderness brings a moment of truth, when we realize how false it is that the only values, moral or artistic or political, are human values. [*1994*]

THROUGHOUT THE 1990S THE COMMUNITY of environmental philosophers grew rapidly across the globe. In 1990 the International Society for Environmental Ethics (ISEE) held its inaugural gathering at a Boston meeting of the American Philosophical Association, the same venue where Rolston had talked with Willard Eddy and secured his job at Colorado State twenty-two years earlier. With a reputation for his philosophical positions as well as his breathtaking work ethic, Rolston was invited to be the founding president of the society and also the editor of its newsletter. This invitation was the start of decades of work he would put into building an international community of environmental ethicists. At this first meeting, the ISEE appointed a joint representative for the United Kingdom and Europe and a second representative for Australia and New Zealand. Within months there were three hundred and fifty members from twenty countries. By 2000 the organization had added representatives for China, Taiwan, Africa, South America, Mexico and Central America,

Pakistan, South Asia, Canada, and Eastern Europe. Environmental philosophy was going global and Rolston was going global with it. Invitations to lecture were now coming in from all over the world.

Rolston rapidly assumed extraordinary international prominence in the field. He lectured at the University of Cape Town in South Africa in 1990 and was a guest of the Institute of Philosophy and the Academy of Social Sciences in China in 1991. He was an official observer in 1992 at the United Nations Conference on Environment and Development in Rio de Janeiro and gave an invited address at the XIX World Congress of Philosophy in Moscow a year later. His environmental philosophy was being read widely overseas, not only in English, but also by translation in Finnish, German, Russian, and a dozen other languages. Within a decade of the publication of his book *Environmental Ethics,* Rolston had lectured in Japan, Brazil, South Africa, Russia, Taiwan, Norway, Finland, Israel, the United Kingdom, Sweden, Denmark, Australia, and Romania. He now made several international speaking trips each year in addition to numerous North American ones. On each visit, he addressed not only university audiences but also professional organizations and government agencies, including groups as diverse as the Wilderness Leadership School in Durban, South Africa, the Society for the Conservation of Nature in Israel, and the Taiwan Forestry Research Institute.

As he was accustomed to do back home, whenever he took a trip overseas for a speaking engagement, he requested that his hosts take him to some notable local park to learn about the ecology of the region. Few philosophers could claim to have seen so many of earth's varied landscapes. He trekked high in the Himalayan mountains, sweltered in the Australian outback, was fed upon by insects in the Amazon forest, baked in the Ethiopian desert, leaned into a fierce wind on top of the Antarctic ice, and stood guard at night against leopards in a South African game park. "Adapted fit," "linguistic molecules," and "systemic value" could be appreciated more or less anywhere in the world. With each new experience Rolston added color and nuance as he articulated his ethic in front of an audience. The freshwater seals he had seen in Lake Baikal had bigger eyes for deeper diving. The Nepalese yaks had especially dense hair to survive the bitter Himalayan winters. As he peppered his seminars with slides and firsthand examples from the South American jungle, the African savannah, and the forests of China, Rolston's classes at Colorado State grew more and more interesting. Yet, appealing as these international dimensions might be to his American audiences, when trying to account for both universality in ethics and local particularity, Rolston found himself running headlong into a serious problem.

For several years now he had been pushing the idea that environmental ethics was about "storied residence." Nonhuman animals had a storied residence in some ecology. A particular historical sequence of events was responsible for making them into *what* they were, *where* they were. Humans, however, were a little different. They had two types of storied residence, an ecological residence and a cultural one. Rolston, for example, was ecologically resident in the Rocky Mountain west, transplanted from the geology and biology of the Shenandoah Valley. Culturally, however, he was an American with Scottish-Irish roots, the son and grandson of Presbyterian pastors, with many of his formative summers spent on Will Long's Alabama farm. His identity resulted from a mixture of both cultural and natural factors. Each of his two types of storied residence influenced how he thought about environmental ethics. He expressed this necessary mix of factors carefully.

> An environmental philosophy does not want merely to abstract out laws and universals, if such there are, from all this drama of life. . . . True, an environmental ethic demands a theory of the whole, an overview of Earth, but not a unity that destroys plurality. We also want an ethic colored by the agents' own history, cultural identification, personal experiences and choices. . . . The moral point of view wants a storied residence in Montana, Utah, Newfoundland, the tall grass prairie, or on the Cape Cod coastline. . . . An ethic has to be instantiated in individuals, who live biographies, each with their local geography.

Honest sentiments designed to show sensitivity to local culture and geography, this mix of unity and plurality turned out to be hard to apply in practice. It was particularly hard to apply when the discussion turned to one of the most important environmental topics—wilderness.

Back in 1967, before contemporary environmental philosophy had arrived on the scene, American historian Roderick Nash pointed out some close connections between the concept of wilderness and a distinctively American type of psychology. Nash suggested that wilderness was a uniquely American idea, the product of a culture fleeing the overcrowded lands of Europe and heading west across a mighty ocean in chase of a dream. Once in the New World, on foot and in wagon trains, the pioneers had for more than a century pushed up against a frontier that was to all appearances both hostile and wild.

The territory Europeans encountered on the North American continent was not just ecologically wild in the sense of being hostile and dangerous; it was also wild in the sense of lacking the established institutional structures of church and crown that the immigrants had left behind. Symbolizing this lack of law to the settlers, American Indians appeared to run wild across the continent, savage and often nearly naked. The land looked unused, untamed, and urgently in need of the imposition of European order. Laboring under the burden of these cultural illusions, the new immigrants set about doing what they thought was God's work, clearing and fencing the land, and turning it to productive use. Entering into the wilderness, the immigrants saw themselves as creating a new promised land. Indigenous people in the way of this mission were either killed or dispatched to reservations where their traditional lifestyles became impossible to sustain.

Two decades before Frederick Jackson Turner declared the western frontier closed in his influential book on the shaping of America, a startling switch began to occur. At the same time as the government was eagerly handing out former indigenous lands to homesteaders, railroad companies, and prospectors for settlement and moneymaking, a few enlightened individuals began suggesting that portions of the new continent should be preserved in public trust. As the western lands became safer for tourists and travelers, visitors from the

East Coast and Europe got a first look at the scenic treasures to be found in the farthest reaches of the new continent. Photographers and artists such as William H. Jackson, Thomas Moran, Albert Bierstadt, and Charles M. Russell produced lithographs, sketches, and paintings that made their way back to the urban population centers. These images and the stories that traveled with them created a stir in Boston, Charleston, and Philadelphia, and as far afield as London and Paris. The reports spoke of unprecedented natural dramas in the remote western landscapes. Some of the stories told about these lands were thought, quite literally, to be beyond belief. Yellowstone basin explorers Charles W. Cook and David E. Folsom were reported to be "so bewildered and astounded at the marvels they beheld, they were, on their return, unwilling to risk their reputations for veracity by a full recital of them." People started talking seriously about preserving parts of unsettled America intact.

As the romantic image of a pristine West filtered back, the idea of wilderness preservation began to register in the American mind. The American landscapes no longer needed to be completely tamed. Some areas could be saved in their wild and natural state. In 1864, New England transcendentalist Henry David Thoreau suggested that portions of the American continent should be preserved primarily "not for idle sport or food, but for inspiration and our own true re-creation." In wildness, as he put it, lay "the preservation of the world." Thoreau had touched on a nascent but powerful cultural sentiment. The call for federal protection of the landscapes reflected a growing American belief in the national significance of wilderness.

Yosemite became the first protected area in the nation when California preserved it "inalienable and for all time" in 1864. In 1872, President Ulysses S. Grant signed the Yellowstone Park Bill into law, creating the nation's—and the world's—first national park. Yosemite and Sequoia followed with their own national park designation in 1890. The preservationist drive began in the West but was not limited to lands beyond the hundredth meridian. New York established the Adirondack State Preserve in 1885, stipulating that it "shall be kept forever as wild forest lands." And in 1891 President Benjamin Harrison created by presidential proclamation the first of the nationwide reserves that would eventually become the 193-million-acre U.S. Forest Service national forest system.

The philosophy driving this preservationist approach lodged itself securely in the American conscience. Uninhabited wild landscapes were valuable to the nation and should be preserved for the public good. Almost a century later, this sentiment became official government policy in the form of the U.S. Wilderness Act. Passed overwhelmingly in both legislative chambers and signed into law by President Lyndon B. Johnson, the 1964 Wilderness Act encapsulated the sentiment Thoreau had articulated a century earlier. The act pointed out how "an increasing population, accompanied by expanding settlement and growing mechanization" threatened to leave no remaining lands in their "natural condition." Preservation of the disappearing wild areas was necessary to ensure for "present and future generations the benefits of an enduring resource of wilderness."

Though the act was ostensibly designed to benefit present and future people, it notably defined wilderness in such a way as to specifically exclude humans. Carefully crafting the language of the legislation, Howard Zahniser wrote that any parcel to be preserved must be

> an area of undeveloped Federal land retaining its primeval character and influence, without permanent improvements or human habitation, which is protected and managed so as to preserve its natural conditions and which . . . generally appears to have been affected primarily by the forces of nature, with the imprint of man's work substantially unnoticeable.

For the first time, the Wilderness Act made it government policy to fence people out of certain parts of America, "where the earth and its community of life are untrammeled by man, where man himself is a visitor who does not remain."

The wild country discovered in North America was cherished

precisely because it represented the antithesis of the overpopulated and overinstitutionalized landscapes Europeans had left behind them. It was an unmistakably immigrant philosophy: wilderness and human inhabitation were deemed incompatible with each other. Indigenous America, of course, had not been consulted.

Twenty-five years after the passage of the Wilderness Act, Rolston found himself to be one of the champions of the philosophy it contained. The Wilderness Society had been one of the first environmental organizations he had joined while living in Virginia in the 1950s. The ethic he now promoted maintained that wilderness was necessarily one of the finest repositories of natural values. Systemic and intrinsic values emerged through the unfolding of natural processes. These processes had been going on for some four and a half billion years, with modern humans arriving on the scene only during the last two hundred thousand or so. Because of the historical reality, Rolston's ethic accorded great significance to events taking place outside of the human sphere. To reach that sphere today, one had to venture into wilderness. There one could find numerous life-forms, from grizzly bears to bull trout, which had been eradicated elsewhere.

As the environmental movement gained strength in America, wilderness preservation became an almost unimpeachable idea. A handful of environmentalists, enlightened by the vocal protests of the American Indian Movement in the 1970s, recognized how distinctively immigrant perceptions had shaped the concept and how it came with devastating consequences for indigenous people. For most environmentalists, wilderness preservation remained widely popular and uncontroversial. And until the end of the 1980s, very few environmentalists ever stopped to consider how America's wilderness idea was being received in parts of the developing world.

In 1989 one of India's foremost public intellectuals shattered the complacency of the North American wilderness movement. Ramachandra Guha was an economist by training but also something of a polymath, highly regarded as an anthropologist, a sociologist, a biographer, and occasional cricket commentator. In the late 1980s, Guha spent two years as a visiting research scholar at the Yale School of Forestry. On his return to India, he wrote a scathing article about radical environmentalism in the United States and its obsession with wilderness protection. The idea that people should be kept out of certain lands for the benefit of wild nature jarred against his cultural intuitions. The obsession with wilderness, he suggested, might be appropriate for a wealthy country with a low population density such as the United States, but it was inappropriate and even dangerous when exported to a highly populated country like India.

Guha pointed out how the settlement histories of America and India differed substantially, the one being a New World country settled and governed by immigrants, the other being among the oldest of inhabited worlds. Apart from the Himalaya and a few patches of untouched forest, almost all Indian landscapes had a continuous and cherished history of human inhabitation. People in rural India tended to live in the same villages as their grandparents, in close concert with land. Technologies were simple and the impact of any one individual on the regional ecology was minimal. America's modern history was a completely different story. The country had been shaped by mass immigration, rapid westward expansion, and a frenzied conversion of the native ecosystems into cities and farms. Historical differences gave Americans and Indians radically different conceptions of what exactly an environmentalist should be protecting.

Guha made it clear that these differences could create insidious effects. American environmentalism, he claimed, was infected by a kind of "biological imperialism." PhD-trained biologists and ecologists, with skills directed toward plants and animals but not people, wielded massive power to determine how government and conservation organizations spent their budgets protecting biodiversity. Beneath their scientific expertise lay little care for the people who would be displaced or the traditions that might be destroyed as a result of their policies.

Nowhere was this arrogance better seen, claimed Guha, than in the efforts of the World Wildlife Fund (WWF) and other large Western nongovernmental agencies "partnering" with the Indian government to save the tiger across its native range. The WWF biologists claimed that incursions by villagers into tiger habitat to hunt and gather firewood were the chief cause of mortality for the endangered populations. Some tigers were killed by villagers in self-defense. Others were poached for tiger parts sold in Asian markets for medicinal purposes. The WWF advocated moving people out of areas where they might easily come into contact with tigers, thus displacing thousands of villagers. The Indian government's Project Tiger mostly acquiesced to these demands.

Indian critics were quick to point out that the people benefiting from the forced relocations were not the locals. Nor were the tigers always better off. The primary beneficiaries were the wealthy foreign elite that could now enjoy ecotour packages through the forest on an elephant's back in the hope of glimpsing one of the rare beasts, never mind the tens of thousands of displaced locals. Guha saw a misdirected Western policy of tigers taking precedence over people. He claimed that, as long-term inhabitants, villagers had every right to stay in tiger country, even if it meant cutting the forest for their needs and killing man-eating tigers when necessary.

Guha's essay, published in the academic journal *Environmental Ethics* and soon after in abbreviated form in the more popular magazine *The New Internationalist,* immediately put the wilderness protection community in North America on the defensive. Though Guha had addressed conservation issues in India, it was obvious that a similar case could be made in many less developed nations. In several African countries, villagers were being fenced out of parks and prevented from protecting their crops against marauding elephants by governments apparently more interested in tourist dollars than the welfare of their own populations. In the Arctic, native villagers were subject to pressures not to hunt seals and whales. Wilderness advocates were soon accused of being "ecofascist" for their propensity to put ecosystem health above all other considerations and "misanthropic" for the little regard they were willing to pay to the well-being of impoverished villagers. Environmentalists looked like the bad guys and Rolston was among them.

In some respects, Rolston was even more vulnerable to the criticism than most. Guha had pointed out the hypocrisy of American environmentalists who did not think twice about traveling overseas or driving halfway across the continent with expensive camping gear to "get back to nature" in a national park. Rolston had recently flown halfway around the world to India, then on to Kathmandu, before taking a helicopter to Lukla in order to see Everest from close up. Observing that the porter carrying his belongings on the trek was too poor to own shoes, Rolston offered to buy him some boots. But the porter told him through a translator that his callused feet would no longer fit inside shoes.

Guha's tiger critique also stung. Rolston had looked for tigers in Royal Chitwan National Park, tramping through the forest on an elephant with local guides in the hope of a fleeting glimpse of the jungle's top predator. Sure, in India and Nepal he had eaten his food off banana leaves, carefully with his right hand only. He was a Christian as well as an environmentalist, sensitive to the locals' religious beliefs and struggles against poverty. Even so, he could never escape the fact that he was indelibly marked as the privileged visitor. When he lectured on habitat preservation in the Himalayas, the well-traveled Presbyterian college professor, raised in Virginia and educated in some of the finest East Coast schools, had only to look down at his expensive hiking boots to feel acutely self-conscious.

Rolston found himself caught up in the middle of what quickly became one of the bitterest debates in the environmental community. The very meaning of wilderness was subjected to a thorough reexamination. Influential books such as Max Oelschlaeger's *Wilderness Idea* and Jack Turner's *Abstract Wild* raised questions about the cultural presumptions inherent in the whole idea of preserving wild landscapes. A series of aggressive critiques of the concept of wilderness ensued. Philosophers and historians questioned the philosophical separation between humans and nature articulated in the Wilderness Act. The critics argued—as Alston Chase had done in Yellowstone—that the idea of land totally devoid of human influence was an ideological fiction. The ecological reality, even on a thinly populated North American continent, was of landscapes continuously influenced by indigenous peoples over millennia.

Another line of thought critiqued wilderness advocates for their apparent desire to create ersatz museums to memorialize the landscapes Europeans had encountered when first colonizing the globe. A good example of this memorializing was the Leopold report's recommendation to maintain—and where necessary recreate—conditions prevailing "when the area was first visited by the white man." Such museums to historical conditions seemed to erase from history the millions of indigenous Americans who had actively managed the landscape with hunting and fire before colonization. Other commentators lamented how the museum approach failed to allow for the dynamic nature of ecological processes. Environmental historian Bill Cronon suggested that wilderness advocates had their eyes on the wrong targets, obsessing about saving small pockets of pristine nature for the recreational elite while neglecting the issues of consumption and inequity threatening these pockets in the first place. By the end of the decade, both sides had enough to say about the wilderness idea for J. Baird Callicott and Michael Nelson to publish a seven-hundred-page collection of essays titled *The Great New Wilderness Debate.* A number of the critics in this book recommended giving up on the idea of wilderness entirely.

In response to the criticisms, wilderness advocates often tended to dig in their heels and insist on the importance of keeping people out of wild areas. One radical environmental group shot back in reply to the likes of Guha and Cronon, "[we are] a wilderness preservation group, not a class struggle group . . . we are biocentrists, not humanists." The arguments became as acrimonious as any in the short history of environmental philosophy. Environmental philosophers in America, who had previously thought of themselves as being on the cutting edge of moral thinking, pushing the boundaries of ethical sensitivity outward into new arenas, were suddenly being portrayed as backward and reactionary. Wilderness preservation went in some people's minds from being the philosophy of a selfless and enlightened group of progressives to being the misanthropic ideology of an extremist fringe. Rolston's position on the intrinsic value of nature's ecosystems put his position dangerously close to the center of the newly vilified group. Though Rolston was never one to put his feelings much on display, the change from being hailed as a progressive champion of the environmental cause to being associated with gross humanitarian injustices must have caused him considerable hurt.

As the criticisms of the wilderness-based approaches to environmental philosophy picked up, Rolston found it incumbent on himself to reply. One possible line of defense was to emphasize his idea of a storied residence in ethics. Cultural and ecological contexts differed considerably from place to place. Rolston knew that priorities had to be different in parts of Africa and Asia. His trips to India and Nepal had opened his eyes to a poverty that went far beyond what he had experienced in rural Appalachia. His trail log from Kathmandu reflected on the conditions the people endured:

> We passed several primitive mills, wheels turned by water, now dry streams. . . . Women carrying hay for cattle, or weaving bamboo baskets, a girl, bare-foot, carrying her baby brother, bare-bottomed. A woman pounding out millet. A woman nursing a child, a man with a bamboo crate of chickens. A profound sense not so much of their being backward as of trekking backward in one's own history, realizing that until the last century or so one's own ancestors lived rather much like this. The rural life, no electricity, no plumbing, no central heating, no motors, no gears, no cars, no paved roads. "By the sweat of thy brow shalt thou eat thy bread." Not so much primitive as primeval. Living antiquity.

Advocates of wilderness preservation, Rolston knew, needed to be aware of the stark human realities. The Colorado high country demanded different management from the Appalachians, and both of those differed from what was appropriate in Guha's India or in Nepal. Rolston agreed that the wilderness idea might need contextualizing before being taken overseas, but this concession to the different needs of different people softened his position only so far. At the end of the day, Rolston still thought it a priority to protect tigers and elephants. The suggestion that the whole idea of wilderness needed rejecting greatly alarmed him.

When Baird Callicott—alongside Rolston one of the most prominent environmental philosophers of this first generation—suggested in an influential paper that the wilderness idea be "revisited," Rolston promptly replied with an article titled "The Wilderness Idea Reaffirmed." He doggedly held on to the original North American conception of wilderness and reiterated its moral significance. He emphasized the importance of protecting areas where the dominant processes were evolutionary and ecological and where humans visited but did not remain. He denied that wilderness advocates sought to preserve static museum pieces, suggesting they were fully ecologically informed and appreciated the dynamism of natural systems. He pointed out that defenders of wilderness were usually active on crucial issues of population and consumption in addition to their obvious interest in saving wildlands for their recreational benefits.

Rolston insisted, much to the irritation of those advocating for indigenous peoples, that any claim of significant modification to North American landscapes before European colonization was wildly exaggerated, often by those desiring to be politically correct and lacking adequate knowledge on the ground. Based on what he had learned about fire ecology, Rolston argued that the impact of intentionally set aboriginal fires was unlikely to be very different from the impact of a natural fire regime. Aboriginal fires depended upon forests and grasses being ready to burn. Indigenous peoples, without a big helping hand from Mother Nature, were unlikely to cause an unnatural ecological transformation. Rolston extrapolated some of this from lessons he learned personally. One weekend in an otherwise wet summer, Rolston was horse packing in Montana's Bob Marshall Wilderness when lightning ignited more than forty fires nearby. Because it was a wet year, most of the fires burned themselves out after consuming a few acres, causing little damage to the forest. In a drier summer six years later, Rolston found himself having to reroute to get out of the same wilderness when the fires joined and burned extensively across an already parched landscape.

Rolston claimed that without wheels and horses, steel axes and saws, aboriginal peoples had never possessed the technologies required to transform natural landscapes on regional scales. In places like Yellowstone that had obviously been subject to indigenous management in the past, Rolston stuck to his position that nature had returned and the human influence had washed out. Portions of the North American landscape, and in particular the North American high country, were still genuinely wild. In those areas, the historical processes dominated and even indigenous peoples were visitors who should not remain.

Rolston was not completely immune to the contention that elements of the wilderness idea were products of the American mind. The lessons he learned at seminary about the historical character of intellectual debates meant he appreciated how ideas were shaped by the history and culture of the people espousing them. At the end of the day, however, he maintained that even when an idea is constructed within a culture, its cultural components need not overwhelm all of its objective reference. Though every idea has some cultural baggage, underneath the baggage something substantial remained.

> Yes, wilderness is, in one sense, a 20th-century construct, as also are the Krebs cycle, DNA, photosynthesis, and plate tectonics. None of these terms were in prescientific vocabularies. Nevertheless, these constructs of the mind enable us to detect what is not in the human mind. . . . There are no doubt many things going on in the wilderness that we yet fail to see, because we do not have the constructs with which to see them. That does not mean, however, that there is no wilderness there, nor that these things are not going on.

Rolston maintained the conviction he had carried with him since high school that science really could give a person a glimpse of what lay outside of the human mind, whether in atoms and galaxies or in DNA and fire ecology. He found himself impatient that the most forceful critiques of wilderness among American scholars were coming from esoteric postmodern thinkers housed in cultural studies and philosophy departments. Refusing to be swayed by the latest academic fashion, he rejected the waves of deconstructive thinking now popular at European and North American universities, retaining his loyalty to the hard realism inherent in the natural sciences. In Rolston's mind, the scientists were much more likely to be good judges of what is real than the cultural theorists. At least the scientists had spent their time in the field getting their hands dirty. To hell with political correctness. In Rolston's eyes, wild nature remained a powerful reality that needed protecting. It did not matter whether it was being protected from the farmers in rural Virginia or the villagers he visited in Nepal or Africa. The argument—and the urgency—was the same.

This debate about wilderness was a defining one for Rolston. He reaffirmed the wilderness idea so strongly not simply because he thought it important to preserve what wilderness remained together with the flora and fauna living there. Even the critics of wilderness generally shared that goal. The reason he fought back so vigorously was that the idea of wild nature's intrinsic value was absolutely central to his whole position in environmental philosophy. The creativity inherent in natural evolutionary processes and passed on in an organism's DNA was the crucial link between biology and ethics in his position. Productive and creative forces were present in wilderness long before humans arrived on the scene. He could not afford to accept the idea that those forces were simply North American cultural constructs. His position was not going to be compromised in order to win any popularity contests.

Roderick Nash had once famously claimed, "Wilderness does not exist. It never has. It is a feeling about a place . . . a state of mind." Rolston felt he could challenge Nash directly on this one. Having been buffered by waves and freezing spray in a zodiac on his way to catch a glimpse of emperor penguins in Antarctica and having been woken by the roar of lions in a Kenyan park, Rolston thought he was entitled to take a different position. "Wilderness is not a state of mind," he replied to Nash, "it is what existed before there were states of mind."

This first part of Rolston's response, reaffirming the wilderness idea in the face of its critics, satisfied almost nobody. Guha's article had created an enormous stir and Rolston appeared to be failing to answer the question of how to care for people in the developing world. The emphasis on wilderness had also put him at odds with a growing impetus for environmentalists to say more about what was going on in urban areas. The tension over the North American concept of wilderness continued unabated, with the charges of ethnocentrism and bias against wilderness advocates appearing to stick. Rolston's participation in the debate had underlined his position as one of the staunchest advocates of the claim that wilderness should be preserved independent of humans. It also lost him some allies among those demanding that American environmentalism should wear a more human face. When he joked to some of his urban colleagues stateside about how Colorado or Montana were better places to be an environmental philosopher than New York or Atlanta, the joke indicated to some of those on the receiving end just what a one-sided conception of environmentalism Rolston had developed.

While reaffirming the wilderness idea in this particular debate, elsewhere in his writing Rolston was pushing what seemed like a conflicting position about human importance, one that even critics such as Guha might have appreciated. In some contexts Rolston was arguing for humanity's elevated moral significance over nature. Humans, he claimed, were the apex of evolutionary achievement and of greater moral significance than anything else in the natural world. Far from being misanthropic, Rolston insisted that humans possessed "the highest per capita intrinsic value of any life-form." He had even gone so far as to suggest there was something vaguely supernatural about *Homo sapiens.*

This was a position you might initially expect one of the most ardent defenders of wilderness to be falling over himself to deny.

11

Feeding People, Saving Nature

> Earth is not just important as our human home. Earth is important because it can be a home planet for all the biota, and that ultimately, is the most important value of all. In the end, we refuse to put humans at the center of concerns. We cannot even put them on top, unless there is something down under; people have to be grounded in their Earth. [*1994*]

TO REACH THE CONCLUSION that humanity stood at the apex of evolutionary achievement, Rolston took a long hard look at the trajectory of the universe since its beginning. An ethic relying so heavily on the idea of storied residence had to be solidly grounded in this historical account. Within the history of the universe, three events appeared to Rolston to stand out. He started calling these events the "three big bangs." The first was the origin of the universe itself, the primordial big bang. Before this event, there was nothing; after it, the universe entered into its long creative journey. The second big bang was the explosion of life on earth approximately ten billion years after the universe's origin. The ability of DNA to transfer information from one generation to the next had resulted in a simply spectacular profusion of life during earth's Cambrian period. The third event Rolston characterized as a big bang was the explosion of mental capacity in the human brain. This final explosion he deemed to be particularly significant.

Although the transmission of information in DNA between generations of nonhuman organisms had made for a fascinating three and a half billion years of earth's history, Rolston argued that with *Homo sapiens* the system had achieved something new. In the brains of modern humans, earth had developed the capacity to transmit information not only on a radically different scale from anything that had come before but also through a radically different method. Information could be transmitted among humans, not only in their genes, but also in their conversations, their books, their libraries, and their databases. "Information in wild nature," Rolston noted, "travels intergenerationally largely on genes; information in culture travels neurally as persons are educated into cultures." With the arrival of *Homo sapiens*, the fundamental process driving earth's biological history forward had changed from one rooted entirely in biochemical mechanisms to one sometimes rooted in cultural ones. In humanity, evolution appeared to surpass itself. This capacity to take information transfer beyond genes was the key indicator of the transcendent human genius.

One of the clearest ways humans demonstrated this genius was by doing science. In science, humans moved beyond the perspective of their biological niche and endeavored to describe the world as if from the outside. The resulting observations were accumulated and passed on through culture as the best available account of the world. Science continually tested the current state of knowledge against reality and educated the next generation to do the same, creating an increasingly vast reservoir of information. Although this cultural reservoir of ideas got its start primarily as a tool for survival under limiting environmental conditions, science took on a life of its own, moving in directions far removed from what was simply necessary for survival and reproduction. Expanding the pool of information became an end in itself. Rolston maintained that no other animal collected and propagated information in this way. In this sense, education for humans was quite literally a "leading out" of nature, allowing them to engage in activities that took them far beyond their biology.

Some of these activities were dazzlingly self-reflective. One such activity by the human brain—itself created by genetic codes—was to decode the very genome responsible for its existence. Another was to reflect on its own workings, coming up with theories of mind and cognition constantly being amended as the science advanced. Not only could humans study the world, they could study *how* they studied the world, marking a double remove from their animal nature. Perhaps most startling of all from an evolutionary point of view, was the ability of *Homo sapiens* to construct norms governing how to behave and then to restrain themselves by acting on the norms. All of these conceptual worlds they created, humans could share and pass on to future generations. No other animal came even close to engaging in such reflective activities. Some started calling this uniqueness the "massive singularity" of humans. Adopting a position that seemed at odds with most environmentalists, Rolston insisted that humans were indeed radically different. In 2003, after receiving repeated criticism for this view, he went to Africa to see if a firsthand encounter with some of our closest biological relatives could persuade him that he was wrong.

If Charles Darwin had dealt the first blow to Christians' comforting prescientific view of human uniqueness, studies of primate DNA appeared to knock it flat. Early studies comparing human and chimpanzee DNA had suggested there was up to a 99 percent convergence between the two genomes. These remarkable findings in molecular biology, coupled with the increasing attention to the fieldwork of primatologists such as Dian Fossey and Jane Goodall, became a staple of environmentalists' views of human relations to the natural world. The knowledge of our biological closeness to whales, chimps, and other higher mammals gave significant argumentative leverage to those rejecting the idea of human privilege deemed responsible for centuries of environmental harm. The environmentalist argument was that it was not humans *versus* animals. Humans *were* animals themselves, and not much different from many of the other creatures with whom we share this earth. A philosopher named David Abram took this line of thinking even further when he started referring to nonhuman organisms as existing in the "more-than-human world," suggesting animals were not merely equal to humans but sometimes superior. A note about our biological kinship with the rest of nature became a standard part of any moral argument to treat the world with more consideration than humans had done in the past.

Rolston had used some of these rhetorical strategies himself when explaining why nonhuman animals had value. Yet he always suspected that this closeness was being exaggerated, not in terms of the DNA numbers but in terms of the closeness that mattered for ethics. The trip to Africa was a counterintuitive mission to confirm not human closeness to nature but human distinctness, a research trip many of his environmentalist friends would find disturbing.

As if to underline the uncomfortable nature of the task, it turned out to be a very difficult trip for Rolston to complete. His first two attempts to see chimps and gorillas in Africa had to be abandoned due to local political unrest. One of these cancellations in 1999 followed the murder of eight tourists in a Ugandan campground ten days before Rolston was due to stay there himself. Rolston—and certainly Jane—began to wonder if the trip was worth it. Perhaps he was destined never to get firsthand confirmation of human uniqueness. Only on the third attempt did his determination to examine humans' closest biological relatives finally pay off.

On his arrival in the Central African forest, Rolston recounted the excitement and adventure in his trail log.

> *Kibale, Uganda.* Up at 5:30 a.m. in total dark, and fumbling to light the kerosene lantern with very poor matches. Breakfast by candlelight. . . . Good chimp track, three knuckles. First we heard one chimp howl. The guide said it was a big male; he was lost and was trying to find the others. Then we found some chimps. I first saw one that was looking straight at me, a small one almost overhead in the dark forest. Two larger ones were hidden in the same tree.

The chimpanzees Rolston encountered in the equatorial forest in Africa quickly confirmed what he had learned in Yellowstone from the wolves; that humans were not the only species to live in close social groups with well-developed social relations. Clearly chimps depended on each other in numerous ways and formed lasting and meaningful bonds. The dexterity and care they showed when grooming each other was shockingly human. Their physical appearance was also a dramatic reminder for Rolston of their close biological kinship. Many of these surface similarities could no doubt be chalked up to the high percentage of shared genetic material.

At the same time as appreciating these obvious commonalities, Rolston also found himself reluctantly noting the vast gulf that remained. Despite the social relations on ample display in Kibale, he found little else in the chimpanzee world that reminded him of the layers of social and cognitive complexity exhibited by humans. Chimpanzees' ability to communicate seemed comparatively undeveloped. Their culture, if you could call it that at all, was primitive. Apart from the odd stick that got poked into a hole, there was an almost complete absence of anything resembling a technology designed to make their lives easier. The chimpanzee brains appeared to offer nothing but the barest precursors of human thought. He saw no evidence of shared conceptual worlds, collaborative learning, or accumulated bodies of knowledge. The chimps were more like a troop of infants than like adult human beings.

Though encountering chimpanzees in the wild gave Rolston a powerful reminder of his close biological kinship with them, it left him with little sense of real psychological similarity. In fact, having traveled to Africa on an airplane using tickets booked over the Internet, and having driven to the forest in Japanese trucks imported by ship to Africa, Rolston was more struck by the huge differences between the abilities of humans and chimpanzees than he was by their similarities. As he had suspected it would, the encounter with chimpanzees seemed to confirm the massive singularity of *Homo sapiens*.

The gorillas Rolston saw in the Bwindi forest a few days later were also in many respects impressive. It was clear Rolston cherished the encounter:

> After about two and a half hours, we began to walk over pushed-down bushes where the gorillas had packed down the bushes to

> make nests. Lots of flies follow the gorillas and the guide said the more the flies, the closer we were. Within one or two minutes of walking I caught a glimpse of a young one. A black patch in the forest moved! We went bushwhacking a bit further and we could see 4–5 moving, resting. The young ones were playing with their mothers. I heard play-chuckle calls. The silverback was lying down and out of sight, more or less, but his huge foot was sticking up in the air.

Rolston was moved by the gorillas' size and their gentleness, and by the way the group functioned as a bonded unit. The huge silverback was a majestic creature by any measure and Rolston admired the gorillas' benign tolerance of his group's incursion. The remarkable experience in the Bwindi forest left him with no doubt that there was a close and humbling kinship between the gorilla species and his own. At the end of the hour spent with the gorillas, however, Rolston found his commitment to the unique abilities of *Homo sapiens* painfully reaffirmed. The apparently basic nature of the communication between the gorillas was the thing that struck him most. The gorillas could grunt to express desires and they touched each other in displays of affection, but Rolston did not leave with the feeling that these behaviors showed much in the way of linguistic or cognitive sophistication. It did not look to him like there was any real awareness of other gorilla minds, any attempts to increase the store of gorilla knowledge, or any efforts to change gorilla opinions. There was no gorilla science going on, no gorilla ethics, and no gorilla religion. He did not see signs of any gorilla genomes being decoded or galaxies being explored. Though gorillas certainly offered a step up in cognitive sophistication from the zebra and antelope Rolston had also encountered in Africa, the gulf between the gorillas and his own species still seemed like a yawning chasm.

Rolston left Africa after his encounter with the gorillas and the chimps more certain than ever of human distinctness. Humans alone were scientists, exchanging theories about the world in which they lived. Only humans were moral agents, able to reflectively discuss and evaluate how they should behave in relation to others. Through their technologies and their complex social networks, humans could relieve the pressures of natural selection on their species in ways quite impossible for chimps and gorillas. Humans used their conceptual worlds to transcend biology in a way that made the rest of nature look quite primitive. Rolston became increasingly insistent on this claim of human distinctness. The lessons of his African encounter solidified into an almost religious affirmation of humanity's uniqueness.

> Animals do not feel ashamed or proud; they do not have angst. They do not get excited about a job well done, pass the buck for failures, have identity crises, or deceive themselves to avoid self-censure. They do not resolve to dissent before an immoral social practice and pay the price of civil disobedience in the hope of reforming their society. They do not say grace at meals. They do not act in love, faith, or freedom, nor are they driven by guilt or seek forgiveness. They do not make confessions of faith. They do not conclude that the world is absurd and go into depression. They do not get lost on a "darkling plain." They do not worry about whether they have souls, or whether these will survive their death. They do not reach poignant moments of truth.

When Rolston saw impoverished Africans choosing to work together to save their wildlife, he saw a paradoxical and heartbreaking type of compassion at work. In the very act of caring for those they cherished as biological kin, humans revealed the "quite stupendous divide that separated us from them."

Much to the dismay of some of his fellow environmentalists, the divide between the human and nonhuman world now occupied a central and unshakable place in Rolston's philosophy. He was not surprised to find the estimates of shared DNA slowly dropping as the science of genomics improved. The latest numbers were indicating something more like 95 percent of DNA shared with chimpanzees, but even this number seemed to Rolston to be somewhat misleading. After all, humans also shared as much as 60 percent of their DNA with fruit flies and a surprising amount with cabbages.

Compared to the chimps, humans possessed a cranial cortex three times as big and several orders of magnitude more powerful. The capacity of their brain set humans apart from any other organism in the living world. Humans alone participated in what Rolston called "cumulative transmissible cultures," passing on information culturally. These mechanisms allowed information transfer to occur infinitely more quickly within culture than it did within genes. This was, for Rolston, the central feature of the human genius. When scientists ventured out to examine the natural world, the most startling phenomenon they encountered was the very brain they used to process their observations. In this respect, humans were unique on earth.

Having worked hard to say how much humans stood out, Rolston was now free to use this commitment to present himself differently in the wilderness debates. The argument about the human mind being qualitatively distinct from anything else in the natural world had at least two different implications for the discussion of wilderness. On the one hand it could be used to support his insistence on the coherence of the wilderness idea. The distinctness of human cognitive capacities could be used to justify the claim that human culture and wild nature differed significantly in kind. If not supernatural, humans were in important ways extranatural. This made it appear reasonable to talk about preserving wilderness and its processes independent of human interference. If, as the U.S. Wilderness Act suggested, land was being preserved to retain its primeval character and influence, then the preservationists were getting it right when they insisted that wilderness was a place where man is "a visitor who shall not remain." Humans were significantly different from anything to be found in wild nature. Their interference really would disrupt wild and spontaneous natural processes.

Rolston's discussion of human exceptionalism not only gave support to the human–nature separation inherent in the wilderness idea, it also went some way toward reassuring antiwilderness critics such as Guha. The uniqueness of *Homo sapiens* ensured that people possessed elevated value. Their astonishing abilities made them, at least in certain regards, a particularly remarkable species. Rolston was insistent on this point: "humans are of the utmost value in the sense that they are the ecosystem's most sophisticated product. They have the highest per capita intrinsic value of any life-form supported by the system."

As the most value-rich product of evolution, *Homo sapiens* warranted special moral consideration. Rolston held, "It is permissible and even morally required to treat unequals with discrimination." This sounded like a guarantee there would be no sacrificing of people for the sake of the tigers. Even impoverished Indian villagers in tiger habitat possessed the human genius Rolston was so keen to highlight.

The problem was whether this pro-villager position would be acceptable to his fellow environmentalists. Rolston, after all, was supposed to be the founder of environmental ethics. Talk of human superiority usually ended up leading to the denigration of all other living forms. If humans really were much more complex, less beholden to their biology, and possessed this special capacity to transmit information neurally rather than genetically, why should they limit their actions in order to protect a natural world that is in many ways inferior to them? Didn't these superior beings deserve dominion? Some philosophers noted with regret that Rolston's commitment to human distinctness seemed more in tune with the traditional Christian view that humans were created in the image of God than with his professed environmentalism.

Here Rolston pushed his conclusions in exactly the opposite direction. His reply to the environmentalists' worry was that with superiority came responsibility. "Different rules do apply to those with superior talents," he argued. Humans did not just have unique value in nature; they also had a unique role. They alone could use their sciences to adopt an overview of what was taking place on earth. This knowledge created a moral obligation to protect the natural values present in the system. Translating superiority as self-importance, he recalled from his Calvinist roots, committed the sin of pride. The possibilities within human nature were stunted if humans acted only out of self-interest. Such was the behavior of animals that possessed myopic and not synoptic views of earth. Species self-interest showed a failure to transcend biology. An elevated understanding demanded

> sentiments directed not simply at one's own species but at other species fitted into biological communities. Interhuman ethics has spent the last two millennia waking up to human dignity . . . environmental ethics invites awakening to a greater story of which humans are a consummate part.

In passages such as these, Rolston thought he had done enough to satisfy both his critics and his followers on the proper place of humans in the natural order.

Despite these considerable efforts to set both humans and their responsibilities apart, Rolston remained haunted by tough questions about how to choose between human and environmental values. The narrow and treacherous path between human and environmental priorities was a challenging one for him to navigate. On a visit with her husband to China, Jane had written to her friends about how the inequities between American and Chinese life bothered her. The tablecloth she had bought at a tourist store had cost more than the annual salary of the Chinese professor hosting their trip.

> Why do I have only to press a switch and in China young and old are loading dried grass or trash on flatbeds behind bicycles to take home to burn for heat? Why are some sorting through garbage piles for items to stuff in drafty holes while I buy their Battenburg lace?

While Jane worried about poverty and issues of international equity, her husband was crisscrossing the Chinese countryside lecturing on the moral importance of protecting panda bears.

When Rolston traveled to Rio de Janeiro for the United Nations Conference on Environment and Development in 1992 he listened to representatives of indigenous people and the global poor vent their frustration at environmentalists telling them not to develop their natural resources. The frustration contained an implicit challenge to Rolston's environmental ethic. It made him realize the need to focus more on issues of human well-being. Humans, he acknowledged, had the highest intrinsic value. Should he not be as focused on their well-being as on the well-being of endangered species? Furthermore, advocates of the global poor pointed out, the two concerns did not have to be antagonistic. The term "environmental justice" had by now been coined, one that conveyed the tight connection that often existed between matters of human and environmental health.

On his return from Brazil to the United States, Rolston made a considerable effort to emphasize the humanitarianism he thought people had missed in his ethic. In an article titled "Environmental Protection and an Equitable International Order: Ethics After the Earth Summit," he argued for justice and equity as necessary partners for environmental protection. He started talking more about the global community taking on issues of overconsumption and underdistribution in the coming century. In 1994 he completed *Conserving Natural Value,* the most policy-oriented of his works to date. In keeping with the general theme of the UN conference, the book reflected an interest in securing freedom from poverty for the third world while also protecting as much of earth's natural value as possible. Each chapter of *Conserving Natural Value* included a number of practical guidelines for how to balance development and preservation. For a while, critics noted, Rolston was paying particular attention to the parts of his ethic showing concern for people over concern for nature. But just when Rolston was starting to improve his humanitarian image, disaster struck. A single article he published in 1996 shattered almost all of the goodwill he had started to accumulate on the people-versus-nature issue.

Years before Ramachandra Guha had ever visited the shores of the United States, environmental ethicists had struggled to find a satisfactory answer to the question of how to balance human interests with environmental obligations. The moral heart of environmentalism, made clear in Rolston's first article in 1975, was a concrete and practical commitment to the nonhuman world. Human behavior needed to be constrained to protect the good of the environment. In the United States, for example, people had been removed from certain areas, sometimes violently, often against their will, in order to create national parks. In these conflicts, a difficult question always remained. How far did this obligation to protect nature go? If it came down to a clear choice between the vital interests of people and the interests of nature, surely it would be wrong not to give priority to people? A major World Health Organization report in 1992 stated these ethical priorities clearly:

> Ensuring human survival is taken as a first-order principle. Respect for nature and control of environmental degradation is a second-order principle, which must be observed unless it conflicts with the first-order principle of meeting survival needs.

In the controversial essay in question, Rolston bluntly denied this claim.

"Feeding People versus Saving Nature" first appeared rather innocuously in a little-read collection on the ethics of world hunger. When people realized what Rolston actually claimed in the essay, reprints started to appear everywhere, both in English and in translation. The essay included the stunning assertion that sometimes we ought to preserve nature at the expense of human life—"Ought we to save nature if this results in people going hungry? If people are dying? Regrettably, sometimes, the answer is yes"—so much for the "massive singularity" earning humans the most intrinsic value on earth. So much for Rolston exhibiting a more humanistic side. All of the efforts he had put into making his position appear more sensitive to people and to social justice were forgotten the moment this provocative essay came into print.

Although it was this particular essay that caused the stir, Rolston had held the controversial position for much longer. He had openly defended the Zimbabwean government's policy of shooting poachers of the highly endangered black rhino on sight. He had visited the game parks himself and spoken with those involved in formulating the management policies. In a taped 1990 debate with Colorado State University colleague and world-renowned animal welfare advocate Bernard Rollin, there had been a celebrated exchange on Rolston's commitment to killing nonnative goats on San Clemente Island off California to protect the endangered plants they destroyed.

Rollin and Rolston where not only philosophically at odds, temperamentally they were leagues apart. Rollin extrapolated Rolston's environmental argument about killing the goats to its logical conclusion:

ROLLIN: On your account I can see no reason for not shooting poachers who are picking an endangered moss.

ROLSTON: Not a bad idea, under certain cases.

ROLLIN: O.K. He's consistent.

ROLSTON: We've got way too many people in the world and too few of certain species. I don't seriously object to shooting people who are poaching rhinoceros.

ROLLIN: No, people who are poaching endangered mosses.

ROLSTON: If you give me a hard enough case in which the only way that I could prevent the elimination of some extremely rare plant was to shoot the would-be poacher—you wouldn't want to be around in the dark, Bernie. . .

ROLLIN: A hard man!

When Rollin and Rolston engaged with each other philosophically, it was sometimes hard to distinguish collegial banter from genuine antagonism. Whether or not the immediate intent of this exchange was comedic, Rolston's claims came across as identical to the radical and politically untenable position known as "ecofascism." Environmental protection, in certain cases, was more important even than human life.

Rolston admitted that the people-versus-nature issue did expose something uncomfortable in an environmentalist's position. In fact, it exposed something so troubling that in his view the question was usually raised for the express purpose of putting environmentalists on the defensive. Neither in the debate with Bernard Rollin nor in the offending article did Rolston back away from the controversial position. Rolston remained firm in his view that there were some occasions when the interests of people, even the vital interests of desperately poor people, should be sacrificed to the interests of nature.

Rolston's article caused an immediate adverse reaction. Anti-environmentalists saw it as a prime example of how green thinkers could go off the deep end. Even those in the environmental community sympathetic to Rolston's general orientation claimed he was wrong on this one. Environmental philosophers Alan Carter, Robin Attfield, Ben Minteer, James Sterba, and Andrew Brennan quickly published articles stressing that Rolston's position was outside the mainstream of environmental philosophy. Brennan in particular decried what he called Rolston's "astonishing misanthropy." Since each of these respondents was an environmentalist by profession, none of them wanted to argue the contrary position that people should always be saved before nature. Rather they suggested that Rolston's construal of the issue—that there was an either/or to be faced between people and nature—was itself part of the problem. The situation was more complex, involving a number of additional social and political factors that needed to be addressed, factors such as the structural dependency between countries, the greed of corporations, and political corruption in the developing world. Most of those responding argued that development done properly would also take care of the environment. Alan Carter's response, "Saving Nature and Feeding People," indicated in its very title the belief that both goals could be achieved simultaneously.

The rash of responses to Rolston's article was in part an attempt at damage control. It was important to reassure those outside environmental philosophy that the discipline had moved beyond the misanthropy of some of its early radical protagonists. Whether or not Rolston himself had moved on, none of these commentators quite knew. Rolston had, after all, suggested that escalating human populations and appetites were "a kind of cancer" on earth. If North America's leading environmental philosopher was a misanthropist and possibly a racist to boot, then the image of the discipline would suffer untold harm. As damage control, the responses seemed to work. Rolston's position against saving people rapidly received almost universal condemnation. Environmental philosophers tried to move quickly on, finding it easier to ignore the fact that their gentlemanly colleague, the former pastor, actually advocated killing (or at least letting people die) in order to protect wildlife.

A closer look at the critical responses left some lingering questions. Rolston had nowhere argued that all environmental problems boiled down to a stark choice between people and nature. Nor did he argue that, when they did, nature should win out every time. The opening pages of Rolston's article pointed out the importance of seeking as many "win-win" situations as possible in people-versus-nature conflicts. He also spent a good deal of time recognizing how political structures often form obstacles that should be addressed before anyone starves in order to save nature. Rolston would have been entirely sympathetic to the suggestion made by one of his critics, Robin Attfield:

> Agonizing about this theoretical question should be replaced by devising policies of development and of preservation in which local people, including rural people, can participate, policies which encourage people both to save wildlife and to feed themselves.

Although Rolston sought win-win solutions as much as anybody else, he elected—perhaps somewhat undiplomatically—to take up the challenge of the difficult cases. If there was no way out of a conflict between people and nature, how should one chose? Directly answering this question was what got him into such hot water. Many of the environmental authors responding to Rolston's paper carefully avoided taking a position on the difficult cases. Rolston noted that they proposed ideal solutions, perhaps possible in some ideal world. The real world, he pointed out, contained the need for tragic choices in real time. Policy makers seldom had the luxury of waiting for the win-win situations to emerge. When Rolston had watched soldiers confiscating cattle in Royal Chitwan National Park in Nepal at the same time as crippled lepers begged him for money and food, he felt like he was learning something firsthand about the hard realities on the ground.

Though it was the dramatic conclusion of Rolston's 1996 article that grabbed all the attention, all the interesting philosophy lay in the careful argument preceding it. Before the controversial conclusion, Rolston had attached a number of careful qualifications to his claim. One of the most important was to insist that his argument would be acceptable only to those living in societies that had already demonstrated an unwillingness to save human lives at all costs. America, he pointed out, was such a society. Expensive meals out at restaurants, lavish Christmas gifts for loved ones, and money spent on college educations could all be redirected toward improving nutrition and health in the third world. Raising the highway speed limit, going frequently to war, and keeping national parks off-limits to settlement were all examples Rolston used to demonstrate how Americans regularly adopted policies that did not prioritize the preservation of life in the ordinary course of events.

Having softened his readers up to their real value preferences, Rolston next argued alongside some of his critics that there were usually several other options for saving human life more effective than sacrificing nature for temporary gain. Reducing national and international disparities in wealth, tackling population issues, addressing corruption, and giving poachers alternative sources of income would all lead to more lasting solutions than sacrificing more of nature. Environmental problems should be fixed in the right places. Stop-gap measures would result in lose-lose situations for both tigers and people in the long term. Deep social challenges would not be resolved by sacrificing earth's remaining wild lands on what Rolston called "the altar of human mistakes."

Finally, in order to identify appropriate occasions on which nature should be saved before people should be fed Rolston provided a screening process. First, he asked policy makers to consider the types of natural values at stake in any one decision and weigh them against the social values to be gained. For example, the value of three thousand surviving rhinos after the African rhino population had been reduced by 97 percent was extremely high. The benefit to the local population trickling down from poaching was extremely low. If the natural values at stake were high enough and the potential human gains low, temporary, or improbable then, in Rolston's view, the sacrifice of nature to save even the poorest people may be immoral.

Rolston was fully aware he was advocating a controversial position. So before he finally drew his conclusion, he carefully restated the list of qualifications:

> 1. If persons widely demonstrate that they value many other worthwhile things over feeding the hungry. . . .
> 2. and if developed countries, to protect what they value, post national boundaries across which the poor may not pass. . . .
> 3. and if there is unequal and unjust distribution of wealth, and if just redistribution to alleviate poverty is refused,
> 4. and if charitable redistribution of justified unequal distribution of wealth is refused,
> 5. and if one fifth of the world continues to consume four fifths of the production of goods and four fifths consumes one fifth,
> 6. and if escalating birthrates continue so that there are no real gains in alleviating poverty, only larger numbers of poor in the next generation,
> 7. and if the natural lands to be sacrificed are likely to be low in productivity,
> 8. and if significant natural values are at stake, including extinctions of species . . . then one ought not always to feed people first, but rather one ought sometimes to save nature.

The former pastor closed his most controversial essay by claiming that God was on his side on this one. In the first endangered species project, God commanded Noah to save earth's species even when people were perishing.

While Rolston was espousing a fairly hard line in his professional life toward third world suffering, it turned out he was acting rather differently in his private life. On a trip to Gandruk in Nepal, writing his trail log under the flickering light of a single forty-watt bulb in the hostel, he was struck by how the water-powered generator proudly showed to him earlier by village elders was the same technology he had encountered over sixty years previously as a small child at the mill in Rockbridge Baths. Appalled by the disparities, he felt a strong Christian obligation to find ways to help. In each of the countries he visited, he looked for worthy local charities. In Nepal, he donated an anesthesia machine to the United Mission Hospital in Tansen. The doctor performing surgery later informed him that his donation had cut postoperative mortality at the hospital by 30 percent. In Bwindi, where Rolston had visited the mountain gorillas, he donated money to purchase forest land for Batwa pygmy villagers to preserve traditional lifestyles. He also donated the annual salary for a doctor in a local clinic. In Kenya, he donated to the Kikuyu Mission Hospital, in South Africa to the Southern African Conservation Education Trust, and in Ethiopia to drought relief projects administered by the Lutheran World Federation. His rule was to try to donate at least as much money to local charitable causes as was spent on his visit. Though Rolston had chosen "saving nature" over "saving people" in his controversial article, behind the scenes he was doing his best to achieve both at once.

By the early years of the new century, Rolston had been at the forefront of environmental philosophy for more than a quarter of a century. Over this time, the central tenets of his view had remained remarkably consistent. The years spent thinking through issues as a rural pastor in Virginia and in his early years as a college professor in the Colorado high country seemed to have given him an unshakable conviction in the framework he now articulated. As Jane well knew, once he had made up his mind on something, her husband rarely looked back. Two and a half decades in the top echelons of academic debates had also hardened him against opposing viewpoints. Even the heartbreaking realities of third world poverty did not fundamentally change how he thought about natural value.

Rolston's longevity in the field and the consistency of his position was by this time making him an increasingly popular target of criticism. Academic philosophers are driven by their training to contest the leading positions. Rolston faced an increasing number of challenges in all the areas he had published. In his environmental aesthetics, his championing of Western science as the key to unlocking appreciation of nature's beauty raised the hackles of those more inclusive of indigenous and Eastern traditions. The fact he ate meat and defended hunting as part of the natural order frustrated those who put a higher priority on animal welfare. Feminist thinkers faulted him for the absence of any gendered analysis of the causes or consequences of environmental destruction. Above all else, his reaffirmation of the wilderness idea and vigorous defense of nature over vital human interests created the strong impression he was cold and insensitive to issues of international poverty. As had sometimes been the case when he was a pastor in rural Virginia, Rolston's position seemed to some of his opponents to be a little too conservative and a little too bound up within his own white Western culture. No one, not even Jane, gained a real sense of how much this mounting academic criticism hurt.

Rolston's philosophical responses to the criticisms were a mixture of careful restatement of his argument and forceful counterpoint to those who still disagreed. Colleagues never saw him display anger, but the strong conviction with which he held his views always came through, occasionally misread as arrogance. Though still a Virginia gentleman in most of his personal interactions, in his professional role he learned how to be quite caustic when he thought it necessary, sometimes dismissive. He frequently accused those who disagreed with him, such as Alston Chase or Baird Callicott, of being "ill-informed" or "lost." One of his favorite rebuttals was to tell his philosophical opponents to "learn some biology." He often damned his critic with false praise by suggesting, "I begin to wonder if my mind is subtle enough to catch the distinction." Just occasionally he brought personal details about his interlocutors' lives, such as their penchant for environmentally destructive air travel or their large family size, into his responses to point out the inconsistencies in the lives of those who criticized him. Occasionally he would bring in personal details of his own—such as the thank-you note from the Yellowstone park supervisor—to provide affirmation of his ideas. Rarely, if ever, did he mention his considerable acts of charity.

Former U.S. Secretary of State Henry Kissinger is reported to have claimed that academic politics are so vicious precisely because the stakes are so small. If Rolston could ever be accused of being vicious in academic debates, it was because he believed, contrary to Kissinger's assertion, that the stakes in environmental philosophy were not at all small. To the contrary, in Rolston's view, simply everything was at stake. And by this, he literally meant everything. Environmental ethics for him was not just about the environment, it was a journey "deeper into ethics than ever before," reaching down to the very roots of what it meant to be human on this diverse and complex planet. The people-versus-nature furor highlighted something Rolston had known since he made the decision to leave his pastorate in Virginia in 1966. Environmental philosophy took a person to the very heart of things: "The creativity within the natural system we inherit, and the values this generates, are the ground of our being, not just the ground under our feet."

Behind everything Rolston wrote in his academic life was his belief that environmental philosophy pointed toward fundamental questions about our lives on earth. It did not simply address moral questions about how to treat plants and animals. It addressed deeper questions of meaning and about how to exist in the cosmos in the fullest sense. Even though he had written a lot of secular articles and books since becoming a college professor, Rolston had never found it possible to do environmental philosophy without his thoughts turning eventually to the divine.

And here the story returns to where it all began.

PART V

Theology for a Green Earth

12

Christianity and the Struggle with Evolution

> Over evolutionary history, something is going on "over the heads" of any and all of the local, individual organisms. More comes from less, again and again. . . . [T]here is a Ground of Information, or an Ambience of Information, otherwise known as God. [*1997*]

BY THE NEW MILLENNIUM, Rolston had achieved the first of the tasks he set himself on leaving the church at Walnut Grove, articulating and bringing to public attention a powerful set of arguments for why humans ought to care about the natural world. He was known globally as an influential and cutting-edge environmental thinker, now often called "the father of environmental ethics." Rolston had lectured to audiences on every continent, including a talk about environmental ethics to the ship's passengers and crew on a voyage to Antarctica. Since 1989, he had appeared annually in *Who's Who in the World.* In 1992 he had been invited to be distinguished lecturer at the twenty-eighth Nobel Conference held in Minnesota. He was listed in a book by Anne Becher as one of three hundred and fifty American environmental leaders from colonial times to the new millennium. The idea he had been propagating of the intrinsic value of nature had appeared in the UN's Global Convention on

Biodiversity and in the Earth Charter. Philosophers and environmental advocates all over the world had studied his work.

So successful had been Rolston's efforts to conform to the academic establishment and write philosophy for a secular audience that it was now possible to read most of his work in environmental ethics and not even know he was a Christian. To achieve the second of the tasks he had set himself on leaving Virginia, Rolston had to persuade Christians that his carefully crafted, secular-sounding arguments for the intrinsic value of nature sat comfortably alongside their belief in the creator God. For the sake of the people he had left behind at High Point Presbyterian thirty-six years ago, Rolston needed to show that modern evolutionary biology could be comfortably reconciled with Christian theology.

The way he saw the issue, not only was it theoretically possible to reconcile evolution and theology, it was also intellectually necessary. Modern science had proven itself to be an extremely effective explanatory tool during the previous two centuries. No religious community could flourish in the long run if it harbored skepticism for the natural sciences. At the same time, it was clear to him that the careful mathematical methods of the sciences left so many phenomena underexplained. Science opened the door to feelings of awe for nature's wonders but those wonders seemed to far surpass mere physical or mechanical explanation. Secular explanations always left a kind of "haunting incompleteness." To fill the gaps, Rolston argued that one needed "science talking to religion and religion talking to science to figure out who we are, where we are, and what we ought to do."

Even though he was sure that a union of biology and theology was essential for a full understanding of our place in the world, the barriers to dialogue between the two approaches seemed almost insurmountable. Some Christians seemed to have long ago given up on the hope of reaching any common ground with evolutionary theorists. To begin the complicated rapprochement, Rolston looked for inspiration from the man most responsible for creating the obstacles in the first place, Charles Robert Darwin.

Quite apart from his reputation as a biologist, Darwin also stood out from many scientists for the amount of thought he put into the problem of reconciling science and religion. Darwin had lived the first two decades of his life as a devout Anglican. In his mid-twenties, after an incomplete attempt to get a medical degree from the University of Edinburgh, he went up to Cambridge to study for holy orders. By the time he enrolled at Cambridge, Darwin was not only a budding cleric but also an accomplished natural historian, an interest that had been with him since his childhood in Shropshire. With the example of Gilbert White in mind, Darwin suspected the life of a country parson would suit both his theological and his biological interests quite well.

Initially, Darwin accepted the story of divine creation told in the Old Testament. When he first read William Paley's *Natural Theology*, a book whose central thesis was that all forms of life were created immutable by God in perfect fit with their neighbors, Darwin was utterly persuaded. Paley saw God as the grand architect of everything in the universe, designing organisms with the meticulousness of a horologist designing watches. Darwin said of Paley's book at the time, "I hardly ever admired a book more . . . I could almost formerly have said it by heart."

After traveling around the world on HMS *Beagle*, Darwin began the long process of losing his faith, precipitated by the growing scientific evidence that the forms of life found on earth were not immutable but had evolved from common ancestors. The conclusions toward which his mind inexorably worked created in him a growing sense of alarm. From the very beginning, he anticipated the trouble these ideas would cause England's Christian establishment. He delayed publication of his most contentious work for many years out of fear of the turmoil he knew it would generate.

The process of turning away from scripture was made particularly wrenching for Darwin by the fact that Emma, his wife, was a lifelong and devout Unitarian. On being informed by her husband of his views on the origins of life, Emma urged Charles to reread the portions of John's gospel telling of the horrible fate of those who turned away from their faith. Emma was terrified at the prospect of eternal damnation for her husband and begged him to reconsider. Despite all the special pleading, the force of the evidence compelled Darwin eventually to reject the creation story of the Old Testament. In the manuscript he later prepared for his family as a memoir, Darwin wrote of his conversion:

> I had gradually come, by this time, to see that the Old Testament, from its manifestly false history of the world, with the Tower of Babel, the rainbow as a sign, etc., etc., and from its attributing to God the feelings of a revengeful tyrant, was no more to be trusted than the sacred books of the Hindoos, or the beliefs of any barbarian.

Wracked with a guilt he compared to confessing a murder, Darwin told his wife he was committed to the view that species were not created by God but descended naturally from a common ancestor.

Though he had rejected the Genesis account of creation, Darwin—like Rolston a century and a half later—knew that a number of the theological questions were still open. Darwin remained attracted to the idea of God lying somewhere behind earth's stunning beauty. Standing in the midst of a Brazilian rainforest in 1833, he rhapsodized, "it is not possible to give an adequate idea of the higher feelings of wonder, admiration, and devotion, which fill and elevate the mind." Evolution by natural selection might be able to explain how species came to be differentiated from each other, but it did not explain how the evolutionary system arose in the first place. There was still room for God as the ultimate creator of the process. In a letter to a colleague in Utrecht written in 1873, Darwin demonstrated that even forty years after he had stood in that rainforest—and fourteen years after he had published *Origin of Species*—he had still not lost the feeling for God's involvement somewhere in the process:

> I may say that the impossibility of conceiving that this grand and wondrous universe, with our conscious selves, arose through chance, seems to me the chief argument for the existence of God; but whether this is an argument of real value, I have never been able to decide.

When Rolston entered the fray more than one hundred years after Darwin's death, Christians were still struggling with exactly the same issue. A natural world of such majesty and complexity suggested to many a divine inspiration. The biological and ecological complexity revealed by the natural sciences in the hundred and forty years since *The Origin of Species* had, if anything, made it even harder to believe that this "grand and wondrous universe" could have arisen through chance alone. At the same time, the ever-increasing weight of scientific evidence had made the theory of evolution by natural selection almost irrefutable. Those wishing to argue from the complexity of the natural world to the involvement of an omnipotent deity needed some more sophisticated strategies than William Paley's nineteenth-century watchmaker God.

Darwin himself had pointed to a simple, but plausible, modification of Paley. In a letter written in 1860 to one of his most enthusiastic correspondents across the Atlantic, a Harvard botanist and Christian named Asa Gray, Darwin confessed that while he no longer believed in the literal creation of earth's species by a benevolent God,

> On the other hand, I cannot anyhow be contented to view this wonderful universe, and especially the nature of man, and to conclude that everything is the result of brute force. I am inclined to look at everything as resulting from designed laws, with the details, whether good or bad, left to the working out of what we may call chance.

God might not have designed every organism as Paley's watchmaker had but he might have set up the physical parameters under which evolution operated.

The precise role of "designed laws" in the cosmos had fascinated Rolston since his studies of physics at Davidson. Physical law was one of the bedrock concepts of science. Much of what happened on earth appeared to happen according to the regular unfolding of physical laws. Within the operation of these laws, his Presbyterian teachers at Davidson had suggested, there was still plenty of room for the involvement of God. Even a universe running like a well-oiled machine prompted questions: How was the system created? Where did the laws governing its operation originate? How did the laws come to have exactly the values they did when their values might have been different?

By the time Rolston's work converged on these topics, modern physics suggested this last question needed an especially good answer. The more scientists had discovered about the universe, the more remarkable it seemed that life had been able to emerge in the first place. Physicists had found that the rate of expansion of the universe after the big bang needed to fall within one part in 10^{55} of its actual value for the universe not to collapse back in on itself. Similarly, if the nuclear weak force had been even incrementally different from its actual value there would have been no hydrogen available. It turned out that the nuclear strong force also had to be within 1.0 percent of its value for the formation of carbon. The sequence of astonishing coincidences continued. If the difference between the mass of a neutron and a proton had not been almost exactly twice the mass of an electron, then either all neutrons would have decayed into protons or all protons would have changed irreversibly into neutrons. If either carbon or oxygen had atomic weights even 0.5 percent different, life would not have been possible. Physics and astronomy turned out to be full of these impossibly small margins for error. As physicist John Leslie had recently noted, the cosmos was "spectacularly fine-tuned for life." Cosmologists proposed the concept of an "anthropic principle" to capture the idea that some force was at work stringing together the unlikely concatenation of laws, physical constants, and historical events that resulted in humankind.

All of this talk about fine-tuning, spectacular coincidences, and remarkable constants had created a growing excitement among theologically minded scientists. Briton Paul Davies put it this way: "Extraordinary physical coincidences and apparently accidental cooperation . . . offer compelling evidence that something is 'going on.'" Stephen Hawking found "clear religious implications." Astronomer Fred Hoyle made explicit what a lot of these physicists were thinking. Somebody, or something, was "monkeying with the physics." As professionals with a good understanding of what counted as evidence, none of these scientists made the claim that the coincidences guaranteed the existence of God. But it was beginning to appear that one could fully buy into the Darwinian account of natural selection and still have room for God lying behind what Darwin had called the "designed laws" of the universe.

During the period that this anthropic principle was gaining currency, Rolston was writing a book titled *Science and Religion.* He could see that the anthropic principle had great appeal to Christians. At one point in his book, Rolston looked as if he bought into it himself. He marveled over how curious it was that the size of the earth sat at the geometrical mean between the size of the known universe and the size of the atom. He also noted that the mass of a human happened to fall at the mean between the mass of the earth and the mass of a proton. Like the physicists he was reading, Rolston was convinced he lived in a universe "mysteriously right for producing life and mind." Science was riddled with extraordinary coincidences that prompted deeper questions about what was really going on.

At the same time as being encouraged by the theological implications of the physics, Rolston could not help noticing that, even if the physics was suggestive, the biology offered no such easy connection to a divinity. Biological history did not appear to be determined in the way that physics was. In fact, biology raised for Rolston some serious questions about whether an anthropic principle at work in the physics could ever guarantee life on earth, let alone guarantee human life.

Rolston was aware that a number of biologists had recently been defending the claim that earth's biochemistry also headed inevitably toward life. Nobel laureate Manfred Eigen insisted that the evolution of life "must be considered an inevitable process despite its indeterminate course." Biochemist and fellow Nobel laureate George Wald was likewise convinced that "the universe breeds life inevitably" as a consequence of natural chemistries. Others were suggesting that evolution headed specifically toward human life, or at least something like it. If one could "play the tape again" in evolutionary history, they said, some sort of intelligent species, if not *Homo sapiens* itself, was predetermined to emerge.

Rolston read with great interest—but also with some suspicion—the work of Simon Conway Morris, a paleontologist made famous by his studies of the Burgess shale. In *Life's Solution: Inevitable Humans in a Lonely Universe*, Morris had argued that evolution settled repeatedly on certain types of solutions to problems. For example, photosensitive organs had evolved multiple times under different conditions, suggesting to Morris that eyes were a solution hardwired into the mechanics of evolution. Wings had developed separately in insects, reptiles, and mammals, indicating that wings might also be one of evolution's inevitable products. Life, Morris observed, "has a peculiar propensity to 'navigate' to rather precise solutions in response to adaptive challenges." Morris called these common solutions "pre-ordained trajectories." Computer modeling had confirmed the convergent tendencies. Morris saw a universe filled with chance and contingency but nevertheless containing definite headings and directions. Earth and its biology were governed by an anthropic principle because humans, or something very like them, were one of the pre-ordained trajectories in the system.

Rolston knew enough about evolutionary biology to agree with Morris that certain solutions to biological problems were more likely than others. Various organisms had clearly converged toward particular forms over evolutionary time. Given enough time in similar environments, marsupials and placental mammals had remarkably similar morphologies despite relatively distinct ancestors. Some were catlike, others doglike. Some predators, some prey. Contrary to Morris, however, Rolston denied that these convergences came anywhere close to bringing inevitability into earth's biological history. Inevitability looked increasingly improbable the further up the phylogenetic scale one traveled. Once the analysis moved over to humans and the products of human culture, the hypothesis of inevitability, Rolston insisted, was beyond credibility.

> There really isn't anything in rocks that suggests the possibility of *Homo sapiens*, much less the American Civil War, or the World Wide

> Web. . . . [T]o say that all these possibilities are lurking there, even though nothing we know about rocks, or carbon atoms, or electrons and protons suggests this is simply to let possibilities float in from nowhere.

The biological version of the anthropic principle looked as if it were a huge act of speculative faith.

The main problem Rolston identified in the attempt to import the anthropic principle into biology was the way it glossed over the inherent unpredictability of biological history. Biological history was not as scripted as these biologists suggested. Maybe it was part of his southern ancestry, but Rolston had always loved how a good story came with numerous twists and turns. Evolutionary history was one of those good stories containing far too much contingency and chance, individuality and eventfulness, to be the preordained unfolding of a divine plan. Evolution was tinkering and makeshift, often wasteful, and frequently filled with extinction and dead ends. The fossil record revealed tales of fantastic creatures created during spectacular explosions of life, followed by cataclysmic events and—for many species—misfortune on a cosmic scale. It was for good reason that people called it the "evolutionary epic."

The reality of earth's history was a constant stream of surprise. With so much chance, it was simply implausible to suggest that God had put the initial conditions in place and then sat back and watched, as the anthropic theorists were claiming. Moreover, in such a world the creator need not have even stuck around to see the unfolding of his plan. He was entirely absent from ongoing biological history. It would make no difference if God were now dead.

Rolston was instinctively uncomfortable with a theology containing such fixedness. It was the same discomfort that had earlier caused him to reject the doctrine of double predestination in Calvinism. The discomfort with predestination was not the only worry. When Rolston had rejected predestination more than four decades ago in his doctoral dissertation in Edinburgh, he had done so in part to allow Calvinists to recover the significance of grace. If everything was scripted so far ahead of time in the natural world, there would be little room for God's ongoing grace. Rolston saw the divine power and the gift of His grace to be much more continuously present in natural history. God did not simply establish the fixed physical laws of the universe and then step away. He had an ongoing engagement with earth's biology. The anthropic principle left Rolston dissatisfied both scientifically and theologically. He started to look beyond the anthropic principle at what other options were on the table.

While some Christian biologists were using sophisticated talk about anthropic principles and preordained trajectories to connect evolution with God, another group was turning back two centuries in the direction of William Paley. In 1802 the theologian Darwin had idolized while studying at Cambridge made a comment that reflected as much about the realities of human psychology as it did about cosmology: "The marks of design are too strong to be got over. Design must have had a designer. That designer must have been a person. That person is GOD."

One of the alternates to the anthropic principle gaining in popularity at the start of the new millennium offered confirmation of William Paley's observation. The theory creating the stir was known as intelligent design. Its theorists did not ignore the scientific evidence in support of evolution by natural selection. Many of them were university-level molecular biologists and hardly in a position to doubt the truth of Darwin. They agreed that in most respects the Darwinian account was correct. But intelligent design theorists insisted that some natural phenomena presented Darwinism with particularly difficult explanatory challenges. As Darwin himself had recognized when contemplating the human eye, certain biological arrangements appear at first to be too complex to have evolved incrementally through random mutation. With some multipartite natural arrangements, it seemed that all the constituent pieces had to be in place for the system to function at all.

The analogy used by the modern intelligent design theorists was the mousetrap. In a mousetrap, the spring, the bottom plate, the attracting food, and the rotating metal bar all need to be present and put together in the proper way for the mousetrap to catch mice.

If any one of the pieces were absent or out of place, the trap would not function at all. The problem this posed for natural selection was that selection worked incrementally, by building up from simpler prior arrangements. It did not allow complex biological phenomena to evolve all at once in their fully functional form.

The intelligent design theorists were essentially pursuing the same line of questioning Darwin had raised about the eye, bringing it to bear on a number of modern biological phenomena. In his 1996 book *Darwin's Black Box*, intelligent design advocate Michael Behe gave the problematic phenomenon the name "irreducible complexity" and offered several examples of the mousetrap phenomenon. One of Behe's examples was the locomotive devices employed by certain bacteria. He claimed the flagella propelling *E. coli* through their aqueous medium would not function if any of their forty or so protein parts were not present. Since evolution by natural selection worked by modifying one piece at a time, the tightly integrated set of proteins should never have been selected. Citing William Paley as a progenitor of his thinking, Behe concluded that biological mechanisms such as the *E. coli* flagella must be designed and implemented by a force outside of natural selection. Evolution was an acceptable theory for explaining most natural forms but it just didn't work for some of the more complex phenomena.

Behe's claims, and those of other intelligent design theorists, prompted a rush of investigative work to disprove the contention that complex mechanisms such as the flagellum of the *E. coli* bacteria, or the mechanism for blood clotting, or certain aspects of the human immune system could not have evolved naturally. A number of biologists, adopting the same tactic Darwin had ended up adopting with the human eye, suggested Behe was simply wrong about these mechanisms being unable to evolve incrementally.

Even though the scientific jury was still out on these issues, Rolston knew Behe had touched on a question he himself thought was important; how to adequately account for the transition from protons and electrons to *Homo sapiens* and the World Wide Web. Was evolutionary biology explanatory enough? Could all of earth's biota have evolved by chance? Did evolution need to be supplemented by something else in order to explain earth's complex and diverse forms? Behe's God was involved in ongoing biological history in a way that the anthropic theorist's God was not, but did Behe's account do any damage to the biology? While Rolston pondered these questions, the intelligent design controversy took on a political life of its own.

A number of overzealous Christians saw intelligent design as an opportunity. The phenomenon of irreducible complexity seemed like a way to redraw some of the lines between science and religion. If certain features of the natural world eluded explanation by the best available scientific theory, then the door was open to an explanatory role for the creator God. To some this also meant the door was also open to putting religion back into the science classroom.

Discussion about whether evolutionary theory was incomplete or inadequate started cropping up at curriculum meetings across America. Various school boards began advocating for the teaching of intelligent design, either as an alternative, or as a necessary supplement, to evolutionary biology. In most cases, the advocates somewhat disingenuously pleaded nothing more than an interest in scientific openness and debate. If Behe was right and Darwin's theory was an inadequate explanation, then the available alternatives, including intelligent design, should be discussed alongside it. This was the opposite of religious dogmatism, they said, it was simply a matter of being open to possible alternatives.

The school board that went furthest and created the most media attention for itself was located in the rural community of Dover in southern Pennsylvania. At a meeting in October of 2004, the board voted to require intelligent design to be discussed as an alternative to evolution in the high school biology classroom. Teachers in the district were instructed to refer their students to the book *Of Pandas and People* as a possible way to plug certain alleged "gaps" in Darwin's theory, a book written by two intelligent design advocates and published by a Christian nonprofit group in Texas specifically for use in the high school biology classroom.

The school board's decision was challenged almost immediately in court by several parents in the Dover district on the grounds that it violated the constitutional separation of church and state. News reports at the time described the battle as "the latest chapter in a long-running debate over the teaching of evolution dating back to the famous 1925 Scopes Monkey Trial." A modern-day media frenzy ensued, exacerbated by a national politics that had been increasingly sympathetic to Christian fundamentalism since the election of George W. Bush in 2000.

United States District Judge John E. Jones ruled in *Kitzmiller vs. Dover Area School District* in December of 2005, siding emphatically with the parents. The judge determined that intelligent design was not a viable scientific alternative. He added to his decision his opinion that parents had been "poorly served" by a school board with a hidden religious agenda. The decision did not make much of a practical difference in Dover. A month before the judge issued his ruling, the school board had been voted out of office en masse by the Dover community, who replaced them with eight members who had pledged to leave intelligent design out of the biology curriculum.

Though Rolston, like Behe, doubted that evolutionary biology was as explanatory as it needed to be, he quickly found he had little sympathy for intelligent design. The idea of God as an engineer building organisms from the ground up was implausible to him for much the same set of reasons he had rejected the anthropic principle. From his theological background, he noted that the word "design" occurred nowhere in the Bible. From his scientific background, he knew that earth's history was too full of eventfulness to talk sensibly of a designer. Phylogenetic histories involved numerous startling and unpredictable turns, making it implausible to say that God had designed any particular mechanism from the start. Evolutionary history displayed contingency and chance, not design, at every turn.

The intelligent design advocates appeared to be caught up in the wrong metaphor. They looked at biological organisms as engineering products with rigorously designed specifications. As Rolston saw it, the engineering metaphor paid too little attention to the fact that every organism had a long and convoluted evolutionary history. The taxonomists and paleobiologists made it clear that there were twists and turns in every species line, requiring God to be more continuously involved. Rolston felt confident the theology was on his side:

> What's wrong with [intelligent design] is not just that it is doubtful science. So far as it is religiously motivated, it is doubtful theology. It makes the creation too passive; the creative relationship between Intelligent Designer and designed product is not relational. It is invasive, manipulative. What results is not Earth bringing forth a self-organizing organism. What results is a marionette.

The God Rolston sought would need to have a much gentler touch in evolution's ongoing and wildly contingent processes.

Rolston was still at a loss for an adequate account of the divine presence in evolutionary history. Anything resembling old-fashioned creationism was long gone from the range of possibilities he was willing to consider. The science simply would not support it. The front-loading of the anthropic principle had also been counted out because it was too deterministic. Intelligent design had now shown itself to require a God who was too manipulative. According to the way Rolston understood the biology, any account of God's participation in ongoing ecological events would have to give due consideration to the openness and chance in the evolutionary story. According to the way he understood the theology, it would also have to give due consideration to the subtle role of God's grace.

Poring again through the latest work by evolutionary biologists, Rolston noted that some of the most seminal minds had begun to identify a list of key directional events in evolutionary history. In their 1995 book, *The Major Transitions in Evolution*, biologists John Maynard Smith and Eörs Szathmáry had pinpointed several transitions that dramatically changed the course of life on earth. Among these were innovations such as life itself emerging on an abiotic planet, the arrival of the eukaryotes from the prokaryotes, the development of multicellular life, and the beginning of meiotic sex. There were other significant moments of transition to be found. The incorporation of mitochondria and then plastids into the cells of eukaryotes, the heat-shock proteins that happened to be transparent and were the precursors of eyes, and the float bladders in fish that became the lungs of terrestrial fauna were similarly important, creative events. Unlike Conway Morris, Maynard Smith and Szathmáry found no reason to suppose these developments were the inevitable results of some fixed biological or physical law. Though the events were compatible with evolutionary theory, they were in no way inevitable or predicted by it. Yet with surprising regularity through earth's history, these transformational events had repeatedly occurred. The regular occurrence of these events caught Rolston's eye. He started calling the sequence of apparent good luck a "cascading serendipity."

One of the things Rolston found curious about these key moments was the way they seemed to involve sideways movements that opened up hitherto unimaginable possibilities. A favorite example for Rolston was the development of mechanisms for hearing sound from the pressure-sensitive cells found in aquatic vertebrates. An apparently unremarkable biological function ended up being coopted for a dramatically different purpose, in the process opening up a vast new arena of possibilities. Originally, the pressure-sensitive skin cells helped a swimming animal orient itself. In their new role they performed a primitive signal transmission function, making it possible for an organism to pick up information from the external environment and transmit it to a processing center in the brain. With this remarkable sideways transition, animals were on the path toward hearing.

Placed in a larger evolutionary context, hearing was an extraordinary development. A consequence of the ability to hear was the possibility of primitive communication. As the ear became more sensitive, boosted in mammals by an amplification mechanism of several small bones and a vibrating membrane, the development of a more sophisticated thorax made the syntax and semantics of advanced language possible. With syntax and semantics came the possibility in humans of passing ideas from mind to mind. A whole new realm of nonbiological information transfer became possible, resulting eventually in the development of culture. The lifting of earth's cybernetic system from its bacterial roots to its human cultural conclusion would not have occurred without this curious cooption of pressure-sensitive cells. Earth's third big bang was made possible by this single improbable development in a fish.

Events such as these set Rolston's theological instincts to wonder. Earthen history had repeatedly exhibited the opening up of dramatic and novel possibilities as a result of such unlikely sideways moves. These serendipitous events, essential for the ultimate generation of complex beings such as *Homo sapiens*, certainly did not have the character of phenomena emerging from front-loaded physical laws or convergent biological solutions. Rolston was convinced that the biological mechanisms of evolution alone could not explain the stream of good luck.

> Looking at a pool of amino acids and seeing dinosaurs or *Homo sapiens* in them is something like looking at a pile of alphabetical letters and seeing *Hamlet*. . . . That is quite as miraculous as walking on water. Something is introducing the order, and, further, something seems to be introducing layer by layer new possibilities of order, new information achieved, not just unfolding the latent order already there from the start in the setup.

Though Behe was off the mark with intelligent design, he was right about the inadequacy of evolutionary theory as a complete explanation.

> If none of the significant trends are derivable from the theory, then the theory is not doing any significant explaining. . . . The universal principle (the best-adapted survive) never entails any particulars (dinosaurs exist), not even when given initial conditions (microbes, trilobites).

Darwin was not wrong, but his theory did not provide enough explanation to make sense of evolutionary history. In order to have an adequate explanation, Rolston felt it necessary to "detect something 'extra' lurking within the serendipitous process." The extra factor took the form of a creativity inherent in the system, an influence that made it possible for evolution to draw order out of the biological phenomena, an influence lying somewhere beyond the biology. Anthropologist Loren Eiseley, writing half a century earlier, had acknowledged the need for an additional explanatory principle working alongside biological history.

> If "dead" matter has reared up this curious landscape of fiddling crickets, song sparrows, and wondering men, it must be plain even to the most devoted materialist that the matter of which he speaks contains amazing, if not dreadful powers, and may not impossibly be, as Hardy has suggested, "but one mask of many worn by the Great Face behind."

The cascade of serendipitous events was beginning to look to Rolston like the mask worn by God, a God who seemed to end up writing remarkably straight with evolutionary history's crooked lines.

13

Cruciform Nature

> To believe in the supernatural is to insist on keeping the concept of the natural open-ended, to refuse to close the system. . . . Aphoristically put, to believe in the holy is the ultimate holism. It insists on the truth, but nothing less than the whole truth, the Holy truth about the forces working for expression in our world and in ourselves. [*1987*]

THE GOD ROLSTON SAW lurking within evolutionary biology did not design particular creatures, nor did he actively manage the details of every probabilistic process. He did not act only at the creation of the universe, nor did he later abandon the natural world to concern himself only with human lives. Rolston proposed a God woven very delicately and continuously into ongoing evolutionary processes. This God did not completely control earth's history. Plenty of contingency remained in the system. Yet his role was to enable, making sure the process as a whole had a positive direction. "Despite the closing down of life lines," Rolston observed, "something seems to open new doors." Although he thought he had found an appropriate place to insert divine inspiration into evolutionary history, Rolston knew he still had plenty of work to do describing what exactly he meant by the phrase "to open new doors."

At the heart of his secular case for nature's intrinsic value had been the conviction that evolutionary epic moved in a positive, value-enhancing direction. Organisms continually faced a struggle against limiting conditions. The struggle made it possible to talk about "achievements" by those that survived and managed to reproduce. New forms of life often established themselves precisely because they could negotiate previously insurmountable problems. DNA locked in the evolutionary upstrokes and passed them on to future generations. Biological wisdom accumulated in the structures and functions of the biota, eventually making possible earth's most remarkable development, the information explosion in *Homo sapiens.* Natural value had steadily increased through time.

This commitment to the idea of values accumulating over time was the key, not only to Rolston's argument for the intrinsic value of nature but also to the argument he now offered for a divine presence. Without accumulating value there was no need to posit a divine force opening new doors and luring life forward through a cascade of creative events. Rolston's philosophy and his theology *needed* a direction to earth's history, and not just any direction, but a progressive, value-enhancing direction. The one problem was that a large number of biologists since Darwin had generally viewed the whole idea of evolutionary progress with intense suspicion.

Darwin himself made a point not to talk of higher and lower life-forms. "After long reflection," he wrote to one correspondent, "I cannot avoid the conviction that no innate tendency to progressive development exists." Modifications occurred randomly. The survival and reproductive advantage offered by a particular mutation determined whether it would be preserved in the species. Changes in environmental conditions meant that modifications made new groups of survivors continually possible. Darwin insisted, however, there was no reason to think of any of these changes having a direction either forward or backward.

Many twentieth-century biologists shared Darwin's view. Stephen J. Gould was particularly dismissive of the idea of a progressive direction to evolution. He characterized *Homo sapiens,* whom Rolston regarded as the most complex and advanced product of evolution, dismissively as "the accidental result of an unplanned process . . . the fragile result of an enormous concatenation of improbabilities, not the predictable product of any definite process." Philosopher of biology Michael Ruse concurred with Gould. "Evolution is going nowhere," he wrote, "and rather slowly at that." These evolutionary theorists insisted that when a species was described as "better adapted" to a particular environment, this did not mean "better" in any progressive sense.

The suggestion that evolution did not move progressively was at odds with what Rolston felt he witnessed in his explorations of the natural world. He had notebooks from all seven continents filled with long lists of the diverse and complex species he had encountered. More important, the idea of progress fit with the numbers offered by the paleontologists. If one measured it in terms of the diversity and complexity of species, Rolston was convinced there was a positive direction to evolution. Over geologic timescales, the evidence in the fossil record appeared to document this direction convincingly.

Rolston knew biodiversity was a complicated concept. It could be measured relative to genes, to species, or to particular geographic environments. But whether counted in marine or terrestrial environments and whether measured in numbers of families, genera, orders, or individual species, overall biological diversity appeared to have been climbing relentlessly since the Archean eon. Three and a half billion years of earth's history governed by Darwinian natural selection had increased the number of life-forms from zero to somewhere between five and one hundred million species today depending on whose estimates you used. This increase was a remarkable achievement in itself. It was even more impressive when considered against the backdrop of half a dozen catastrophic extinction periods, each of which inflicted heavy setbacks on the biodiversity accumulated up to that point. In his commitment to the idea of increasing biodiversity Rolston found a prominent ally in Harvard biologist E. O. Wilson. Wilson shared Rolston's assessment that the increase in species diversity counted as progress:

> Progress, then, is a property of the evolution of life as a whole by almost any conceivable intuitive standard. . . . In spite of major and

> minor temporary declines along the way, in spite of the nearly complete turnover of species, genera, and families on repeated occasions, the trend in biodiversity has been consistently upward.

Not only had there been an upward trend in biodiversity, Rolston also thought it clear there had been an accompanying trend upward in complexity. Like diversity, complexity was not a straightforward concept, and its rise had not been universal across the biota. Some groups, such as the arthropods, appeared to have gained only limited complexity in millions of years. In others, such as the marine species living in the ocean depths that had lost the ability to see, complexity had gone downward. Yet if complexity was measured relative to the ability to gather and process information, earth's history on the whole seemed to have exhibited a clear rise over evolutionary time. In primates particularly, there had been tremendous advances in capacities for motor control, manipulation of objects, sentience, communication, and acquired learning, with the gains ultimately leading to the creation of human culture.

Perhaps unsurprisingly, given his commitment to *Homo sapiens* being the pinnacle of evolutionary achievement, Rolston insisted that the most startling demonstration of complexity in the natural world was the modern human brain, an organ that appeared on earth only in the very recent geologic past. A little neuroscience revealed some breathtaking numbers. The brain was estimated to contain approximately 10^{11} neurons. Each neuron had hundreds or even thousands of synaptic connections to other neurons and to nerves. If all of the connecting fibers of the brain were stretched out in a thin line, they would circle the globe forty times. The postsynaptic membrane through which these connections operated contained over a thousand different proteins in the signal-receiving surface, making it one of the most complex molecular structures in the body. In addition to the sheer volume of the connecting surfaces, each neuron could operate with at least ten different levels of electrochemical activity. Based on these figures, the number of possible activation states of the brain exceeded the number of atoms in the known universe. For all practical purposes, the creativity inherent in the human brain was without limit.

Remembering the statistic that the mass of a human stood about midway between the mass of a proton and the mass of the earth, Rolston noted how complexity appeared to rise to an extraordinary apex at the planet's middle range. "We humans do not live at the range of the infinitely small, nor at that of the infinitely large," he remarked, "but we may well live at the range of the infinitely complex." Life on earth had started with prokaryotes and now included humans with their remarkable brains. Rolston struggled to see how Darwin, had he known these numbers, would have failed to view this as progress. Rolston aligned himself instead with geneticist Theodosius Dobzhansky:

> Seen in retrospect, evolution as a whole doubtless had a general direction, from simple to complex . . . to greater and greater autonomy of individuals, greater and greater development of sense organs and nervous systems conveying and processing information about the state of the organism's surroundings, and finally greater and greater consciousness.

Going simply by the numbers, if diversity and complexity of the biota were the measure, Rolston was completely convinced that there was a genuinely progressive direction to evolutionary history.

If evolution had a progressive direction, questions of how and why inevitably arose. Trying to account for evolutionary progress was the point where the explanations, for Rolston, turned religious. Biology alone was not enough to explain the astonishing history. Even if complexity and diversity had nowhere to go but up after the first single-celled organism, there was no reason inherent in the biology for the trend to continue in the spectacular fashion it did. Explaining earth's trajectory as a product of randomness was really to offer no explanation at all. Mutations in genomes sometimes took complexity downward. Among all the species that had ever existed, 99 percent were now extinct. Rolston saw here not just the opportunity, but also the necessity, of adding an additional explanatory layer to make sense of the generally upward trend. The way he read the evolutionary epic, earth needed a God to "lure protozoans into persons" over three billion years. The challenge he faced was to figure out just the right mix of biology and theology to simultaneously satisfy both his Darwinian and his Christian friends.

First the biological component. Rolston saw mutating genetic sets functioning through history as something akin to a search program. Continually evolving genomes generated trial-and-error solutions to problems set by the environment. Genetic mutation, crossover, drift, allelic variation, cutting and splicing, insertions, and deletions all allowed the different individuals within a species to explore the possibility spaces. Genomic behavior amounted to a kind of "generate and test" facility. The genomes that showed they could solve the problems by finding the hospitable spaces, or simply by diverging from other forms within a single space, got passed on. This was earth's evolutionary mechanics at work, a cybernetic system inherently structured to explore possibilities.

Next the theological component. When he had stood as a child at the side of the Maury River and watched the hydraulic pump doing work for the local farmers, Rolston had learned it was possible to move water uphill against gravity's downward pull. Unlikely possibility spaces lay hidden in the hydrodynamics, waiting to be explored. The movement toward increasing order and complexity in the biota seemed to demonstrate a parallel tendency in natural history; life somehow moving uphill against the stream of death and disorder one would expect from nature's entropic tendencies. Earth clearly had no mechanical pump and Rolston had learned enough from the intelligent design debates to know that talk of a divine engineer or designer was problematic. Nevertheless something seemed to be there ensuring the possibility of this countercurrent. Something made a probabilistic process run reliably in an unlikely direction. "Some force is present," he said, "that sucks order in superseding steps out of disorder." The cascading serendipity of those extraordinary and innovative sideways moves suggested to Rolston a divine power at work.

In place of the farmer's pump countering the downward movement of the river's energy, Rolston posited a theological "Ground of Information" opening up the possibility spaces and presenting options to the genetic search program. Nothing in the physics and chemistry of life determined these possibilities and Rolston thought it disingenuous to let them "float in from nowhere." These exploratory possibilities need an originating source and, for Rolston, that source was the divine. "God is the atmosphere of possibilities," he claimed, "the metaphysical environment in, with, and under first the natural and later also the cultural environment."

Being very careful not to put God into the mechanics of the process, Rolston suggested his presence lurked within evolution's randomness and creativity, introducing the new possibility spaces for a searching genome. On balance, the divine force acted as an "orderly innovating principle," slipping information into the world and presenting options in such a way that natural selection tended, especially in the more complex lines, progressively upward toward its apex in humans. Darwinian evolution was the correct analysis of what was going on physically and biochemically, but the Darwinian account was embedded in a larger theological story about creative possibilities continually being made available. God was not an architect, a designer, an engineer, or a mathematician, but he seemed to be active in earth's history as an encouraging presence, an inspiration, a catalyst, and a presenter of possibilities.

Rolston worked hard to keep an adequate distance between his view and the views of those who embraced intelligent design or the anthropic principle. By talking about God as presenting an "atmosphere of possibilities" he hoped to retain all the integrity of the science and at the same time show there was still room for an ongoing divine role in evolutionary history. To keep the science and theology in a productive tension required an extremely delicate balancing act. The balancing act was necessary, however, because the biology alone, to his mind, could never be explanatory enough.

> One can say, restricted to science and with maximum economy, that only the environment acts to shape the organism. But one can say,

> religiously and with maximum insight into the meaning of what is taking place, that this formative force, which elicits so much from so little, mind from life, life from matter, is a divine force in the world.

All this was very theoretical on the surface, but the idea of God as an encouraging presence and a presenter of possibilities had a particular resonance at this juncture in Rolston's own life. At about the time these thoughts were crystallizing in his mind, Rolston and his wife were shepherding their two children, Giles and Shonny, through the difficult teenage years. Giles, his hiking companion for many years, had started getting himself into trouble. As Rolston struggled through the challenges of parenting, he found his role to be a revealing reflection of God's more subtle role in biological history. The traditional Christian metaphor of God the father seemed to fit particularly well with how Rolston saw evolutionary biology. God kept up a gentle pressure on turbulent processes that could always go astray. Throughout earth's storied history, God "lured," "coaxed," "shaped," and "nurtured" for diversity and complexity, just as parents try to do the same to develop wisdom and maturity in their children. God opened doors and introduced possibilities, some of which the evolving biota had explored, some of which they ignored. God did not compel anything, but without God opening the doors and presenting the good options, the possibility of progress would have been slight. This was God the father at work in the evolutionary story as much as in the human one.

The idea of God the father giving direction to the evolutionary process was not the only part of the account that Rolston found theologically satisfying. Given the Calvinist commitments rooted deep within him, another appealing aspect of his proposal was that it gave special significance to the operation of God's grace. It was grace that made possible the cascade of serendipitous events over the system's history. Grace ensured the stepping through the doors to explore the evolutionary possibilities. Grace was ultimately responsible for earth's progressive history. Not all of the possibility spaces got explored, but the ones that were had on balance propelled diversity and complexity dramatically upward. None of this happened by necessity and none of it happened easily. There was struggle and often great suffering involved for individual organisms in the process. Nevertheless, within the constant interplay of struggle and grace lay the basis of earth's progressive history.

The interplay between struggle and grace was not just a feature of evolutionary biology; it was also a feature of human life. For the former pastor, this parallel brought all the abstract reflections in evolutionary biology firmly back in touch with the everyday lives of Christians. As a lover of family lore, Rolston knew of numerous occasions when his own ancestors had struggled against misfortune while benefiting from God's grace. On his mother's side, there had been Daniel Long, the great-great-grandfather who lost a daughter to the infected overcoat. There was also Daniel Long's grandson, Will, who had been raised in poverty in Depression-era Alabama. Both of these ancestors made it through these hardships with the help of their faith and God's grace. The story of family suffering that touched Rolston the most, however, was the story embodied by the tombstone he made a point of visiting every few years in New Providence Cemetery whenever he was back in the Shenandoah Valley.

On graduation from Hampden-Sydney Seminary in the mid-1890s, Rolston's paternal grandfather, Holmes Rolston I, had been called to his first pastorate in Horton, West Virginia. There he had worked among the hardened men of the Allegheny Mountain lumber camps. The former cowboy took with him to Horton his new wife, Jacqueline Campbell. In keeping with Virginia lore, the couple had married just after the goldenrod bloomed in September 1895, in a ceremony presided over by the bride's brother-in-law.

Life in the lumber camps of West Virginia was harsh. There were no churches yet established in that part of the county. The lumbermen were constantly drunk and fighting. The wooden shack in which the couple lived was cold in winter and full of flies in summer. Despite the hardships, the young pastor and his wife devoted themselves to each other and to their work creating a church community.

In late summer they celebrated the birth of their first child, Archibald Campbell Rolston. The celebration was brief. Archibald was sickly, barely clinging to life from the moment of his birth. In the oppressive August heat the infant never seemed to gain much strength. The lumber camp was a long way from any hospital and his parents watched helplessly as the color of their son's cheeks paled. Archibald died in the Horton home less than three weeks after his birth.

Determined the infant should have a proper burial, Rolston's grandfather put his son's body in the back of a one-horse buggy and set out on in the early morning for his brother-in-law's New Providence Church. He rode alone for seventy miles, stopping only to rest and water the horse. Upon arrival at New Providence, the pastor found his anguish magnified. His sister-in-law's own firstborn, a young girl of two weeks named Jacqueline Campbell Wilson, had died the day before his own child. The two Campbell sisters had each lost their firstborn within twenty-four hours. The infants were buried side by side in the New Providence cemetery with a single tombstone recording the double tragedy. Intense suffering seemed to be part of every family story. Acceptance of this suffering as part of God's order seemed to be a demand of faith.

Rolston knew that all Christians must face the challenge of theodicy, the challenge to reconcile the acute suffering so clearly present in so many lives with the belief in an all-powerful, all-knowing, and all-loving God. The challenge was even more acute for Rolston and for other Christians who knew the natural order in the kind of detail he did. They not only had to justify how God could allow children to die from illnesses, nations to fight violent conflicts against each other, and earthquakes to wreak havoc over innocent citizens. They also had to accept that God would create a system in which millions of trout smolts hatched for the sake of only a few surviving into adulthood, male grizzly bears sometimes killed and ate their own cubs, and wildebeests on the African savannah got ripped apart by a lion's jaws while still alive. The problem of evil in the human world seemed to be replicated a thousand times over in the natural world.

Part of what had driven Charles Darwin from his faith was his despair at the amount of suffering, misery, and waste the supposedly benign creator had permitted on earth. Having lost his mother to a tumor when he was a child and his favorite daughter, Annie, to a painful illness when she was only nine, Darwin had cause far beyond the morphology of the Galapagos finches to embrace natural selection over divine creation. If there were a divine creator, Darwin once suggested, he would have to be a better God than the one responsible for this earth. In a letter to his Christian friend in America, Asa Gray, Darwin cited the actions of parasitic wasps and playful cats as just two examples of the catalog of horrors he saw within creation.

> I had no intention to write atheistically. But I own that I cannot see, as plainly as others do, and as I should wish to do, evidence of design & beneficence on all sides of us. There seems to me too much misery in the world. I cannot persuade myself that a beneficent and omnipotent God would have designedly created the Ichneumonidae with the express intention of their feeding within the living bodies of caterpillars or that a cat should play with mice.

Darwin framed with directness the theological question that ultimately defeated him: "what advantage can there be in the sufferings of millions of the lower animals through almost endless time?" Rolston knew this question was in need of a persuasive theological answer.

Rolston was in a better place than most to reflect on the problem of evil. As the son and grandson of a pastor and then as a pastor himself, Rolston had spent a good portion of his life in regular contact with human anguish. Reflecting on these experiences, Rolston had written:

> A pastor shares the fortunes and misfortunes of his people, their virtues and vices, cradle to the grave. . . . Farmers struggled to make crops, to stay out of debt. Fathers struggled to make ends meet, mothers to keep marriages going, children and youth to stay in school and out of trouble. Death often took an early toll: alcoholism, highway accidents, or disease, a son killed in War.

Such events demanded constant efforts to explain to grieving relatives why their suffering was not evidence against the existence of a good God.

Rolston got a start on his theodicy with a lesson about suffering embedded deep in his southern ancestry. One of the truths he had learned from growing up in the South was that a great loss can sometimes result in a great gain. Southerners, he knew, had ultimately benefited from losing the Civil War. The South was wrong about slavery. Even though hundreds of thousands of Confederate soldiers, including some of Rolston's own ancestors, had died on behalf of a mistaken conviction, the defeat eventually made it possible for the South to become a better, more prosperous, and more harmonious region. The initial loss was entirely necessary for the later gain.

To Darwin's key question about the purpose of all the suffering, Rolston expanded on the theodicy offered by second-century theologian Irenaeus of Lyon. Irenaeus had pointed out how much there was to be gained from suffering. One had only to look past the immediate hardship to see the later spiritual gain. Suffering was "soul-making," providing opportunities to develop the Christian virtues. Rolston appreciated how one could extend Irenaeus's observations to the natural world. Where Darwin had seen only dark clouds in the parasitic wasps and predatory cats, Rolston found an awe-inspiring silver lining.

Pain and suffering in nature, Rolston argued, were not an end in themselves but a means of protecting and enriching life. Their biological function was not simply the immediate reaction they generated in an organism, the spitting out of some foul-tasting berries or the fighting back against a predator. In addition to causing immediate alarm and distress, suffering in the natural world ultimately led to an improved ability to respond to environmental threats. An organism better prepared to navigate a threat was more likely to survive to reproduce. The drive to avoid suffering and death ensured that the divinely presented possibility spaces continued to get explored.

> Adversities make life go and grow. . . . The cougar's fang has carved the limbs of the fleet-footed deer. . . . The brown fat cell evolves because of the need to survive the fierce wintry cold. Such a milieu of struggle might seem to suppress life, but there is a paradoxical reverse result—that life is disciplined, strengthened, and improved.

Earth accumulated and enriched its natural values not just in spite of death and suffering but in many ways because of it. In the larger evolutionary scheme of things, suffering was woven deeply into the workings of a prolific and remarkably generative system. The system gathered and retained biological information in part as a response to the continuous and often painful deaths of millions of individuals. Death and suffering were almost never the last word. The innovative possibility spaces Rolston suggested were opened up by God required the suffering if they were ever to get explored.

Darwin thought God should have created a better world. Rolston accepted the existing world as the one required for creative evolutionary achievement.

> To be alive is to have problems. Things can go wrong just because they can also go right. In an open, developmental, ecological system, no other way is possible. . . . Death can be meaningfully put into the biological processes as the necessary counterpart to the advancing of life. . . . The evolutionary sequence . . . seems to thrive on the tragic accidents that slay all the successive individuals.

The enlivening truth Rolston had learned over hundreds of days exploring the high country, during countless hours of counseling his parishioners, and through listening to numerous family stories about the hardships faithfully endured by his southern ancestors was that life survives, and even flourishes, in the midst of its perpetual perishing. This insight became the basis of Rolston's theodicy for both the human and the natural world.

For visual confirmation of life surviving amidst perpetual perishing, Rolston found no greater assurance than in the irenic spring bloom of the pasqueflower. Every March in Colorado, Rolston sought out a glimpse of the year's first pasqueflowers pushing their way through melting alpine meadows at the first hint of spring warmth. Like generations before him, Rolston saw profound meaning in this vernal rebirth. Pasqueflowers in the Colorado high country were particularly potent symbols of life's annual triumph over death. The petals were purple or violet, colors associated with Lent and Jesus' suffering for human sin. The flowers blossomed around Easter, the time of the crucifixion and resurrection. Rolston reflected on the lessons embodied by the pasqueflower:

> In its annual renewal as the first spirited flowering against the blasts of winter, it is a sign against the eternal storm . . . when it bursts forth with the breaking up of the raging winter . . . it is such a reminder of life's survival, indeed of a prospering such that it hopes for paradise. . . . This pasqueflower endures the winter in noble beauty; but its suffering is not only the shadow of its beauty, it is among the roots that nourish it.

The role of suffering and death in the natural world as a precursor to greater florescence mirrored a similar process in culture. "Life is suffering," Rolston acknowledged, "but life is suffering through to something higher." Both in nature and in culture, the constant tension between suffering and grace was a value-enhancer. The urge to avoid suffering had generated countless innovations in cultures across the world. The switch toward living in mutually supportive social groups, the development of agriculture, continual improvements in technology and medicine were all propelled by the drive to avoid suffering. The scriptures were not wrong when they warned that "We eat our bread with tears." The nearby shadow of suffering and death pushed people to explore the possibility spaces constantly opening up on earth. In *Homo sapiens*, at the apex of biological complexity, this pressure had yielded spectacular results.

Rolston recognized that suffering seemed to become more acute the further one moved up the evolutionary scale. "Each seeming advance" he warned, "steps up the pain." As neuronal complexity increased, so did the capacity to suffer and so did the depth of the pain. The increase in pain, however, turned out to be incredibly productive. Moving up the phylogenetic scale, suffering did not just propel better adaptation and fit, it opened the doors to entirely new phenomena. Sentient creatures with advanced cognitive capacities developed increasingly sophisticated responses in the face of suffering, learning the ability to love, to desire, and to care. The ability of

Homo sapiens to be self-reflective and empathetic made it possible to value the welfare of others, opening doors into the moral world. As Darwin had recognized, suffering required not just explanations that gave it biological function but also explanations that gave it meaning. Rolston saw how suffering caused people to ask profound questions about their lives. It offered them the opportunity to love. In many cases, it caused them to become religious.

Suffering and death, first appearing to be negative features of earth's story, ended up being responsible for generating a great deal of the positive values on earth. Christianity had recognized this from the start.

> To be chosen by God is not to be protected from suffering. It is a call to suffer and to be delivered as one passes through it. . . . So far from making the world absurd, suffering is a key to the whole. . . . The capacity to suffer through to joy is a supreme emergent and an essence of Christianity. Yet the whole evolutionary upslope is a lesser calling of this kind, in which renewed life comes by blasting the old. Life is gathered up in the midst of its throes, a blessed tragedy, lived in grace through a besetting storm.

These facts of both biology and culture provided Rolston with his central theoretical bridge between evolution and theology.

With this significance placed on suffering and grace, it became clear how all of Rolston's philosophy was in the end rooted firmly within his ancestral Christian faith. The capacity to suffer through to joy was the supreme essence of Christianity. After nearly forty years of reflecting on these issues, he now held in his grasp an account of evolution that might appeal to his former parishioners at High Point. These parishioners knew that Christ lived a life of sacrificial suffering so that humanity could reach something higher. Rolston could tell them that, just as Christ's death on the cross was necessary for the possibility of human redemption from sin, so were the deaths of millions of animals through evolutionary time necessary to drive the trajectory of earth's history upward. This process generated immense value. The parallel between the suffering of Christ and the evolutionary process led Rolston to describe natural processes as "cruciform." Nature replayed the passion of Jesus and in the process created more diverse and complex life-forms. The suffering was redemptive. In this parallel between nature and culture, Rolston saw not just moral but also religious value. Evolution was not hostile to Christian belief; it offered as clear a demonstration imaginable of the message contained in the life of Jesus. *Via naturae est via crucis.* The way of nature is the way of the cross.

Cruciform nature, lives of redemptive suffering, and histories of accumulating values were the links between Rolston's work in environmental philosophy and his work in theology. Returning as always to his Calvinist convictions, Rolston found himself constantly in awe of the appearance of God's grace in all this eventfulness. Nothing in the geologic history made life and mind inevitable. Persons did not have to emerge from protozoa. The suffering on earth did not have to yield such spectacular results. The fact that something made it so was an indication to Rolston of the gift of God's grace.

> There is startling fertility, genesis. This is among the best established facts. This creativity we inherit, and the values this generates, are the ground of our being, not just the ground under our feet. Nature is grace, whatever more grace may also be. If anything at all on Earth is sacred, it must be the enthralling creativity evident on our home planet.

Gazing out upon cruciform nature, Rolston compared his perspective to that of Annie Dillard in her classic work of nonfiction, *Pilgrim at Tinker Creek.* Dillard, a biologist by training, had recognized the impressive generativity of earth but, like Darwin, sought in vain to make peace with all the suffering. She confessed she was ultimately unable to find that resolution. Dillard found it impossible to see the earth as the sort of place a benevolent God would have created: "I came from the world, I crawled out of a sea of amino acids, and now I must whirl around and shake my fist at that sea and cry 'Shame!'" The misery on display in both the natural and the human world left Dillard embittered and angry.

Natural processes never tempted Rolston to shake a fist in disgust. Earth's history may have read like a passion play containing plenty of reason for despair and grief, but within the story of suffering lay ample evidence of divine grace and numerous opportunities for redemptive love. Rolston discovered in the evolutionary epic not something to resent, but something to revere. He offered his reply to Dillard and her feelings of anguish.

> If I were Aphrodite, rising from the sea, I think I would turn back to reflect on that event and raise both hands and cheer. And if I came to realize that my rising out of the misty seas involved a long struggle of life renewed in the midst of its perpetual perishing, I might well fall to my knees in praise.

Here in cruciform nature, with life consistently suffering through to something higher, Holmes Rolston III found evolutionary biology and Christian theology convincingly making their peace.

Epilogue

AT 5:45 A.M. EACH MORNING, the bedside alarm goes off at Rolston's modest ranch-style house on a quiet residential circle in Fort Collins, Colorado. He rises quietly and moves to the basement, where he does some early morning stretches. Rolston fixes his breakfast, usually just a big bowl of cereal, and reads journal articles while he eats. By a little after 6:30 a.m., he is sitting in his office at a desk that looks out onto a neatly kept suburban backyard. A few family photographs hang on the wall, a memento or two from trips overseas, and a number of shelves filled with a variety of books and journals, stacked neatly but not excessively so. Rolston switches on the computer and, as the machine beeps and starts to boot up, picks up the notes and correspondence he had set on his desk the previous day. He adjusts his glasses and, as he does every morning at this time, begins his day of work.

For most of his career, Rolston remained unsure of the impact of his work. Receiving an unimpressive brown envelope in the mail one day and thinking he was getting another solicitation for money from the University of Edinburgh, his alma mater, Rolston threw the letter into the trash before retrieving it, and getting one of the biggest shocks of his life. This 1997 invitation to give the Gifford Lectures in Edinburgh was recognition of the magnitude of Rolston's scholarly achievements in natural theology. His receipt of the Templeton Prize six years later was a different kind of honor, but one similarly indicative of his international repute.

The Templeton Prize rewards those who make "progress or discoveries about spiritual realities." No other environmental philosopher has ever won the prize. Perhaps in part because of the size of the award—an amount always pitched deliberately above the Nobel and MacArthur awards to make it the largest in the world—the Templeton Prize has occasionally generated controversy. Critics question the idea that one can really talk about "discoveries" in religion, at least in the way one can talk about discoveries in science. Given the emphasis Rolston has maintained throughout his career on firsthand experience of scientific nature, few theologians or philosophers can more appropriately claim their work has represented discovery than Rolston.

The same day that Rolston received the prize, he donated all the money to endow a chair in science and religion at the college where he earned his undergraduate degree in physics, Davidson in North Carolina. When he handed over the check to a Davidson representative shortly after the award ceremony in Buckingham Palace, he announced in his soft southern cadence that being a millionaire for five hours had "felt pretty good." Having his story told on the radio and in hundreds of newspapers for a few days seemed to leave him genuinely perplexed for a moment about whether to feel self-important or profoundly embarrassed. He had, after all, set out to understand the world. Had he really managed in some small way to change it?

Rolston devoted his working life to bringing the biological sciences, Christian theology, and philosophical ethics into contact. It is fair to say he succeeded in weaving them together more tightly than anyone has done before. Certain themes anchor the framework. The idea of "survival amidst perpetual perishing" is central. "Suffering through to something higher" is a crucial common feature of both human and natural communities. The interplay of law and chance, the notions of narrative and storied residence, and the ethically laden terms "value" and "achievement" bring a tight coherence to the position normally elusive in such complex territory. One of Rolston's successes has been to keep these questions of meaning in the closest possible contact with the explanations offered by the natural sciences. Helping give birth to the field of environmental ethics is possibly the best-known of Rolston's intellectual achievements. Reducing the tensions between evolutionary theory and Christian theology is an equally remarkable one. God stays in the picture but without the need for creationism, intelligent design, or a heavily deterministic system.

When Rolston encountered Albert Einstein at Princeton in the 1950s, he reflected on the great physicist's claim that the deepest mystery of all was the earth's comprehensibility. Rolston recognizes the relationship between mystery and intelligibility to be the tension that drives every investigation embarked upon by the human mind. Toward the conclusion of one of his books, Rolston quotes the theoretical physicist Victor Weisskopf describing the knowledge we do accrue as "an island in the infinite ocean of the unknown, and the larger this island grows, the more extended are its boundaries toward the unknown."

Rolston is well aware that a great deal of work remains to be done even within his own position. How God opens up the possibility spaces in earthen history is unclear. Doubts about whether earth needs to contain so much pain and suffering remain. The exact degree of open-endedness in biological history still presents a challenge. While the main components of Rolston's view are cemented firmly in place, the nuances will require some tinkering for as long as the boundaries of science continue to press outward. To his colleagues and friends, Rolston has never shown doubt about the religious convictions at the center of his work. There are questions, however, that still bother him. He asks in genuine puzzlement about some parts of Africa, "How can they be Christian for so long, and yet remain so poor?" Such painful uncertainties, he knows, keep constant company with both science and religion. Rolston insists that no other species but humankind is capable of reflecting on these great unknowns and no other species feels compelled by its very essence to do so.

Not everyone is convinced that Rolston has articulated the most compelling environmental philosophy. In the last few years, environmental pragmatists have argued that Rolston set the field off in the wrong direction. A core group of urban environmentalists critique what they think is his pro-wilderness bias. Secular Greens don't like the religious elements that linger in the subtext. Consistent with his taciturn nature, Rolston suffers these criticisms silently and with dignity. He knows that defending ideas, like defending life-forms, is a constant struggle.

The man who has already inscribed the epitaph Philosopher Gone Wild onto his gravestone in the family plot at Hebron Presbyterian is a grandfather now. He wonders how much of his grandchildren's lives he will live to see. Though he officially retired in 2008, he is still active at Colorado State University, a mentor to numerous students and colleagues. His seminars are filled, as they have been for forty years, with students who have come from across the country and the world to study with the "father of environmental ethics." He has not lost his birthright as a soft-talking and genial southern gentleman but he still generally prefers to spend his time studying alone rather than in company. He and Jane return to Virginia when they can, taking advantage of any opportunity that comes their way to be in the Shenandoah Valley in springtime. Rolston continues to work incessantly, seven days a week, reviewing articles, writing recommendations, preparing the presentations he gives all over the country. On the weekends, he still does philosophy in the Colorado high country, bringing home the odd leaf and flower to key out with his reference books back home. He moves more slowly over the rocky paths now, less able and perhaps also less interested in reaching the summits. In the woods, he reminds himself, as he has done for years, "you're not going somewhere; you are already there."

Each spring around the vernal equinox, Rolston makes a solitary pilgrimage into the mountains in search of the purple bloom of the pasqueflower. When he sees some petals emerging alongside a snowbank, he pauses for a long moment of reflection, finding reassurance in life's survival after a winter of perishing cold.

> For longer than we can remember flowers have been flung up to argue against the forces of violence and death, because that is what they do in and of themselves, and thus they serve as so ready a sign for any who encounter them in a pensive mood, wearied of the winter, frightened by the storm, saddened by death. This is why it is liberating to find the pasqueflower bearing with beauty the winds of March.

The pilgrimage is more than a habit. It is an affirmation. The rebirth of the pasqueflower at winter's end remains for Rolston the most fitting symbol of the conviction that resides at the heart of his thinking.

> The way of nature is, in this deep though earthen sense, the Way of the Cross. Light shines in the darkness that does not overcome it.

Select Bibliography of Rolston's Work

BOOKS

Conserving Natural Value. New York: Columbia University Press, 1994.

Environmental Ethics: Values in and Duties to the Natural World. Philadelphia: Temple University Press, 1988.

Genes, Genesis and God: Values and Their Origins in Natural and Human History. New York: Cambridge University Press, 1999. Gifford Lectures, University of Edinburgh, 1997–1998.

Philosophy Gone Wild. Buffalo, NY: Prometheus Books, 1986.

Science and Religion—A Critical Survey. Philadelphia: Templeton Foundation Press, 2006. First published 1987 by Random House.

ARTICLES AND CHAPTERS

"Are Values in Nature Subjective or Objective?" *Environmental Ethics* 4 (1982): 125–151.

"Biology and Philosophy in Yellowstone." *Biology and Philosophy* 5 (1990): 241–258.

"Caring for Nature: From Fact to Value, from Respect to Reverence." *Zygon: Journal of Religion and Science* 39 (2004): 277–302.

"Challenges in Environmental Ethics." In *Environmental Philosophy: From Animal Rights to Radical Ecology*, ed. Michael E. Zimmerman, J. Baird Callicott, Karen J. Warren, Irene J. Klaver, and John Clark,

82–102. 4th ed. Upper Saddle River, NJ: Pearson/Prentice-Hall, 2005. First published 1993 by Prentice-Hall.

"Creation: God and Endangered Species." In *Biodiversity and Landscapes,* ed. Ke Chung Kim and Robert D. Weaver, 47–60. New York: Cambridge University Press, 1994.

"Disvalues in Nature." *The Monist* 75 (1992): 250–278.

"Does Nature Need to Be Redeemed?" *Zygon: Journal of Religion and Science* 29 (1994): 205–229.

"Duties to Endangered Species." *BioScience* 35 (1985): 718–726.

"Environmental Ethics: Values in and Duties to the Natural World." In *Ecology, Economics, Ethics: The Broken Circle,* ed. F. Herbert Bormann and Stephen R. Kellert, 23–96. New Haven: Yale University Press, 1991.

"Feeding People versus Saving Nature." In *World Hunger and Morality,* ed. William Aiden and Hugh LaFollette, 248–267. 2nd ed. Englewood Cliffs, NJ: Prentice-Hall, 1996.

"Generating Life on Earth: Five Looming Questions." In *The Evolution of Rationality,* ed. F. LeRon Schults, 195–223. Grand Rapids, MI: Eerdmans, 2006.

"Genes, Brains, Minds: The Human Complex." In *Soul, Mind, Brain: New Directions in the Study of Religion and Brain-Mind Science,* ed. Kelly Bulkeley, 10–35. New York: Palgrave Macmillan, 2005.

"Is There an Ecological Ethics?" *Ethics* 85 (1975): 93–109.

"Mountain Majesties above Fruited Plains: Culture, Nature, and Rocky Mountain Aesthetics." *Environmental Ethics* 30 (2008): 3–20.

"Naturalizing and Systematizing Evil." In *Is Nature Ever Evil? Religion, Science and Value,* ed. Willem B. Drees, 67–86. London: Routledge, 2003.

"Naturalizing Values: Organisms and Species." In *Environmental Ethics: Readings in Theory and Application,* ed. Louis P. Pojman and Paul Pojman, 107–120. 5th ed. Belmont, CA: Thomson Wadsworth, 2008.

"Nature and Culture in Environmental Ethics." In *Ethics: The Proceedings of the Twentieth World Congress of Philosophy,* ed. Klaus Brinkmann, 151–158. Bowling Green, OH: Philosophy Documentation Center, 1999.

"Nature for Real: Is Nature a Social Construct?" In *The Philosophy of the Environment,* ed. Timothy D. J. Chappell, 38–64. Edinburgh: University of Edinburgh Press, 1997.

"The Pasqueflower." *Natural History* 88, no. 4 (April 1979): 6–16.

"Value in Nature and the Nature of Value." In *Philosophy and the Natural Environment,* ed. Robin Attfield and Andrew Belsey, 13–30. Cambridge: Cambridge University Press, 1994.

"Values Gone Wild." *Inquiry* 26 (1983): 181–207.

"What Is a Gene? From Molecules to Metaphysics." *Theoretical Medicine and Bioethics* 27 (2006): 471–497.

"The Wilderness Idea Reaffirmed." *Environmental Professional* 13 (1991): 370–377.

"Wildlife and Wildlands: A Christian Perspective." In *After Nature's Revolt: Eco-justice and Theology,* ed. Dieter T. Hessel, 122–143. Minneapolis: Fortress Press, 1992.

FESTSCHRIFT

Preston, Christopher J., and Wayne Ouderkirk, eds. *Nature, Value, Duty: Life on Earth with Holmes Rolston, III.* Dordrecht: Springer, 2007.

A full bibliography is on the Rolston Web site, often with downloadable materials.

http://lamar.colostate.edu/~rolston

Extensive Rolston archives, including media, are in the Colorado State University Library.

Index